T0180766

Advances in Intelligent Systems and Computing

Volume 535

Series editor

Janusz Kacprzyk, Polish Academy of Sciences, Warsaw, Poland
e-mail: kacprzyk@ibspan.waw.pl

About this Series

The series "Advances in Intelligent Systems and Computing" contains publications on theory, applications, and design methods of Intelligent Systems and Intelligent Computing. Virtually all disciplines such as engineering, natural sciences, computer and information science, ICT, economics, business, e-commerce, environment, healthcare, life science are covered. The list of topics spans all the areas of modern intelligent systems and computing.

The publications within "Advances in Intelligent Systems and Computing" are primarily textbooks and proceedings of important conferences, symposia and congresses. They cover significant recent developments in the field, both of a foundational and applicable character. An important characteristic feature of the series is the short publication time and world-wide distribution. This permits a rapid and broad dissemination of research results.

More information about this series at http://www.springer.com/series/11156

Jeng-Shyang Pan · Václav Snášel
Tien-Wen Sung · Xiao Dong Wang
Editors

Intelligent Data Analysis and Applications

Proceedings of the Third Euro-China
Conference on Intelligent Data Analysis
and Applications, ECC 2016

 Springer

Editors

Jeng-Shyang Pan
College of Information Science
  and Engineering
Fujian University of Technology
Fuzhou, Fujian
China

Tien-Wen Sung
College of Information Science
  and Engineering
Fujian University of Technology
Fuzhou
China

Václav Snášel
Faculty of Electrical Engineering
  and Computer Science
VŠB-Technical University of Ostrava
Ostrava-Poruba, Moravskoslezsky
Czech Republic

Xiao Dong Wang
College of Information Science
  and Engineering
Fujian University of Technology
Fuzhou
China

ISSN 2194-5357   ISSN 2194-5365 (electronic)
Advances in Intelligent Systems and Computing
ISBN 978-3-319-48498-3   ISBN 978-3-319-48499-0 (eBook)
DOI 10.1007/978-3-319-48499-0

Library of Congress Control Number: 2016954940

This Springer imprint is published by Springer Nature
The registered company is Springer International Publishing AG
The registered company address is: Gewerbestrasse 11, 6330 Cham, Switzerland

Preface

This volume of *Advances in Intelligent Systems and Computing* contains the accepted papers presented in the ECC 2016, the Third Euro-China Conference on Intelligent Data Analysis and Applications, which was held in Fuzhou City, China, during November 7–9, 2016. The aim of ECC is to provide an internationally respected forum for scientific research in the broad area of intelligent data analysis, computational intelligence, signal processing, and all associated applications of AIs.

The third edition of ECC was organized jointly by Fujian University of Technology and Fujian Provincial Key Laboratory of Big Data Mining and Applications, China, and VSB-Technical University of Ostrava, Czech Republic. The conference was co-sponsored by Taiwan Association for Web Intelligence Consortium and Immersion Co., Ltd.

The organization of the ECC 2016 conference was entirely voluntary. The review process required an enormous effort from the members of the international technical program committee, and we would therefore like to thank all its members for their contributions to the success of this conference. We would like to express our sincere thanks to the host of ECC 2016, Fujian University of Technology in China, and to the publisher, Springer, for their hard work and support in organizing the conference. Finally, we would like to thank all the authors for their high-quality contributions. The friendly and welcoming attitude of conference supporters and contributors made this event a success!

September 2016

Jen-Shyan Pan
Václav Snášel
Tien-Wen Sung
Xiao Dong Wang

Organization

Honorary Chair

Xinhua Jiang   Fujian University of Technology, China

Advisory Committee Chairs

XiaoDong Wang   Fujian University of Technology, China
KeShou Wu   Xiamen University of Technology, China
Bin-Yih Liao   Kaohsiung University of Applied Sciences,
            Taiwan

Conference Chairs

Jeng-Shyang Pan   Fujian University of Technology, China
Vaclav Snasel   VSB-Technical University of Ostrava,
            Czech Republic
Luo Hao   Zhejiang University, China

Program Committee Chairs

RiQing Chen   Fujian Agriculture and Forestry University, China
Muhammad Khurram Khan King Saud University, Saudi Arabia
Yi Wu   Fujian Normal University, China

Invited Session Chairs

Tien-Wen Sung   Fujian University of Technology, China
Tarek Gaber   Suez Canal University, Egypt
Tsu-Yang Wu   Harbin Institute of Technology Shenzhen
                           Graduate School, China

Local Organizing Chairs

FuMin Zou   Fujian University of Technology, China
Zhiming Cai   Fujian University of Technology, China

Electronic Media Chair

Tien-Wen Sung   Fujian University of Technology, China

Publication Chairs

Xiangwen Liao   Fuzhou University, China
Pei-Wei Tsai   Fujian University of Technology, China

Finance Chair

Hong Chen   Fujian University of Technology, China

International Program Committee

Aarti singh   Maharishi Markandeshwar University, India
Abdel hamid Bouchachia   University of Klagenfurt, Austria
Abdelhameed Ibrahim   Mansoura University, Egypt
AbdElrahman Shabayek   Suez Canal University, Egypt
Abd. Samad Hasan Basari   Universiti Teknikal Malaysia Melaka, Malaysia
Abraham Duarte   Universidad Rey Juan Carlos, Spain
Ahmed Anter   BeniSuef University, Egypt
Akira Asano   Kansai University, Japan
Alaa Tharwat   Suez Canal University, Egypt
Alberto Alvarez   European Centre for Soft Computing, Spain
Alberto Cano   University of Cordoba, Spain
Alberto Fernandez   Universidad de Jaen, Spain
Alberto Bugarin   University of Santiago de Compostela, Spain
Alex James   Indian Institute of Information Technology
                           and Management - Kerala, India

Alexandru Floares   Romania
Alma Gomez   University of Vigo, Spain
Amelia Zafra Gomez   University of Cordoba, Spain
Amira S. Ashour   Tanta University, Egypt
Amparo Fuster-Sabater   Institute of Applied Physics (C.S.I.C.), Spain
Ana Lorena   Federal University of ABC, Brazil
Anazida Zainal   Universiti Teknologi Malaysia, Malaysia
Andre Carvalho   University of Sao Paulo, Brazil
Andreas Koenig   Technische Universitat Kaiserslautern, Germany
Anna Bartkowiak   University of Wroclaw, Poland
Anna Fanelli   Universita di Bari, Italy
Antonio Peregrin   University of Huelva, Spain
Antonio J. Tallon-Ballesteros University of Seville, Spain
Anusuriya Devaraju   Forschungszentrum Julich GmbH
Aranzazu Jurio   Universidad Publica de Navarra, Spain
Ashish Umre   University of Sussex, UK
Ashraf Saad   Armstrong Atlantic State University, USA
Ayeley Tchangani   University Toulouse III, France
Aymeric Histace   Universite Cergy-Pontoise, France
Azah Kamilah Muda   Universiti Teknikal Malaysia Melaka, Malaysia
Bartosz Krawczyk   Politechnika Wroclawska, Poland
Beatriz Pontes   University of Seville, Spain
Brijesh Verma   Central Queensland University, Australia
Bing-Huang Chen   Fujian University of Technology, China
Carlos Barranco   Pablo de Olavide University, Spain
Carlos Cano   University of Granada, Spain
Carlos Fernandes   GeNeura Team, Spain
Carlos Garcia-Martinez   University of Cordoba, Spain
Carlos Lopezmolina   Universidad Publica de Navarra, Spain
Carlos Morell   Universidad Central Marta Abreu de Las Villas,
                               Cuba
Cesar Hervas-Martinez   University of Cordoba, Spain
Chang-Shing Lee   National University of Tainan, Taiwan
Chao-Chun Chen   Southern Taiwan University, Taiwan
Chia-Feng Juang   National Chung-Hsing University, Taiwan
Chin-Chen Chang   Feng Chia University, Taiwan
Chris Cornelis   Ghent University, Belgium
Chun-Wei Lin   Harbin Institute of Technology Shenzhen
                               Graduate School, China
Chuan-Kang Ting   National Chung Cheng University, Taiwan
Chuan-Yu Chang   National Yunlin University of Science
                               and Technology, Taiwan
Chu-Hsing Lin   Tunghai University, Taiwan
Coral del Val   University of Granada, Spain

Crina Grosan	Norwegian University of Science and Technology, Norway
Cristina Rubio-Escudero	University of Sevilla, Spain
Cristobal Romero	University of Cordoba, Spain
Cristobal J. Carmona	University of Jaen, Spain
Chia-Hung Wang	Fujian University of Technology, China
Chia-Jung Lee	Fujian University of Technology, China
Dalia Kriksciuniene	Vilnius University, Lithuania
David Becerra-Alonso	ETEA-INSA, Spain
Detlef Seese	Karlsruhe Institute of Technology (KIT), Germany
Edurne Barrenechea	Universidad Publica de Navarra, Spain
Eiji Uchino	Yamaguchi University, Japan
Eliska Ochodkova	VSB-Technical University of Ostrava, Czech Republic
Elizabeth Goldbarg	Federal University of Rio Grande do Norte, Brazil
Emaliana Kasmuri	Universiti Teknikal Malaysia Melaka, Malaysia
Enrique Herrera-Viedma	University of Granada, Spain
Enrique Yeguas	University of Cordoba, Spain
Eulalia Szmidt	Systems Research Institute Polish Academy of Sciences, Poland
Eva Gibaja	University of Cordoba, Spain
Federico Divina	Pablo de Olavide University, Spain
Fernando Bobillo	University of Zaragoza, Spain
Fernando Delaprieta	University of Salamanca, Spain
Fernando Gomide	University of Campinas, Brazil
Fernando Jimenez	University of Murcia, Spain
Francesc J. Ferri	Universitat de Valencia, Spain
Francesco Marcelloni	University of Pisa, Italy
Francisco Fernandez Navarro	University of Cordoba, Spain
Francisco Herrera	University of Granada, Spain
Francisco Martinez-Alvarez	Pablo de Olavide University, Spain
Francisco Martinez-Estudillo	University Loyola Andalucia, Spain
Frank Klawonn	University of Applied Sciences Baunschweig, Germany
Gabriel Luque	University of Malaga, Spain
Gede Pramudya	Universiti Teknikal Malaysia Melaka, Malaysia
Giacomo Fiumara	University of Messina, Italy
Giovanna Castellano	Universita di Bari, Italy
Giovanni Acampora	University of Salerno, Italy
Girijesh Prasad	University of Ulster, UK
Gladys Castillo	University of Aveiro, Portugal
Gloria Bordogna	CNR IDPA, Italy
Gregg Vesonder	AT&T Labs Research, USA
Huiyu Zhou	Queen's University Belfast, UK

Hai-Yan Yang	Fujian University of Technology, China
Ilkka Havukkala	Intellectual Property Office of New Zealand, New Zealand
Imre Lendak	University of Novi Sad, Serbia
Intan Ermahani A. Jalil	Universiti Teknikal Malaysia Melaka, Malaysia
Isabel Nunes	UNL/FCT, Portugal
Isabel S. Jesus	Instituto Superior de Engenharia do Porto, Portugal
Ivan Garcia-Magarino	Universidad a Distancia de Madrid, Spain
Jae Oh	Syracuse University, USA
Jan Martinovic	VSB-Technical University of Ostrava, Czech Republic
Jan Plato	VSB-Technical University of Ostrava, Czech Republic
Javier Sedano	Technological Institute of Castilla y Leon, Spain
Javier Perez	University of Salamanca, Spain
Jesus Alcala-Fdez	University of Granada, Spain
Jesus Serrano-Guerrero	University of Castilla-La Mancha, Spain
Jitender S. Deogun	University of Nebraska, USA
Joaquin Lopez Fernandez	University of Vigo, Spain
Jorge Nunez Mc Leod	Institute of C.E.D.I.A.C, Argentina
Jose Valente De Oliveira	Universidade do Algarve, Portugal
Jose Luis Perez de la Cruz	University of Malaga, Spain
Jose Villar	Oviedo University, Spain
Jose M. Merigo	University of Barcelona, Spain
Jose-Maria Luna	University of Cordoba, Spain
Jose Pena	Universidad Politecnica de Madrid, Spain
Jose Raul Romero	University of Cordoba, Spain
Jose Tenreiro Machado	Instituto Superior de Engenharia do Porto, Portugal
Juan Botia	Universidad de Murcia, Spain
Juan Gomez-Romero	Universidad Carlos III de Madrid, Spain
Juan Vidal	Universidade de Santiago de Compostela, Spain
Juan J. Flores	Universidad Michoacana de San Nicolas de Hidalgo, Mexico
Juan-Luis Olmo	University of Cordoba, Spain
Julio Cesar Nievola	Pontificia Universidade Catolica do Parana, Brazil
Jun Zhang	Waseda University, Japan
Jyh-Horng Chou	National Kaohsiung First Univ. of Science and Technology, Taiwan
Jerzy W. Rozenblit	University of Arizona, USA
Kang Tai	Nanyang Technological University, Singapore
Kaori Yoshida	Kyushu Institute of Technology, Japan
Kazumi Nakamatsu	University of Hyogo, Japan
Kebin Jia	Beijing University of Technology, China

Kelvin Lau	University of York, UK
Kubilay Ecerkale	Turkish Air Force Academy, Turkey
Kumudha Raimond	Karunya University, India
Kun Ma	University of Jinan, China
Leandro Coelho	Pontificia Universidade Catolica do Parana, Brazil
Lee Chang-Yong	Kongju National University, Korea
Leida Li	University of Mining and Technology, China
Leon Wang	National University of Kaohsiung, Taiwan
Liang Zhao	University of Sao Paulo, Brazil
Liliana Ironi	IMATI-CNR, Italy
Lincoln faria	Universidade Federal Fluminense, Brazil
Luciano Stefanini	"University of Urbino" "Carlo Bo", Italy
Ludwig Simone	North Dakota State University, USA
Luigi Troiano	University of Sannio, Italy
Luka Eciolaza	European Centre for Soft Computing, Spain
Liang-Cheng Shiu	National Pingtung University, Taiwan
Macarena Espinilla Estevez	Universidad de Jaen, Spain
Manuel Grana	University of Basque Country, Spain
Manuel Lama	Universidade de Santiago de Compostela, Spain
Manuel Mucientes	University of Santiago de Compostela, Spain
Marco Cococcioni	University of Pisa, Italy
Maria Nicoletti	Federal University of Sao Carlos, Brazil
Maria Torsello	Universita di Bari, Italy
Maria Jose Del Jesus	Universidad de Jaen, Spain
Mariantonietta Noemi La Polla	IIT-CNR, Italy
Maria Teresa Lamata	University of Granada, Spain
Mario Giovanni C.A. Cimino	University of Pisa, Italy
Mario Koeppen	Kyushu Institute of Technology, Japan
Martine De Cock	Ghent University, Belgium
Michael Blumenstein	Griffith University, Australia
Michal Kratky	VSB-Technical University of Ostrava, Czech Republic
Michal Wozniak	Wroclaw University of Technology, Poland
Michela Antonelli	University of Pisa, Italy
Mikel Galar	Universidad Publica de Navarra, Spain
Milos Kudelka	VSB-Technical University of Ostrava, Czech Republic
Min Wu	Oracle, USA
Mohamed Eltoukhy	Suez Canal University, Egypt
Mohamed Khairy	Suez Canal University, Egypt
Mohamed Tahoun	Suez Canal University, Egypt
Mona Solyman	Cairo University, Egypt

Nilanjan Dey   Techno India College of Technology, India
Noor Azilah Muda   Universiti Teknikal Malaysia Melaka, Malaysia
Norberto Diaz-Diaz   Pablo de Olavide University, Spain
Norton Gonzalez   University of Fortaleza, Brazil
Noura Semary   Menofia University, Egypt
Nurulakmar Emran   Universiti Teknikal Malaysia Melaka, Malaysia
Olgierd Unold   Wroclaw University of Technology, Poland
Oscar Castillo   Tijuana Institute of Technology, Mexico
Ovidio Salvetti   ISTI-CNR, Italy
Ozgur Koray Sahingoz   Turkish Air Force Academy, Turkey
Pablo Villacorta   University of Granada, Spain
Patrick Siarry   Universit de Paris, France
Paulo Carrasco   Universidade do Algarve, Portugal
Paulo Moura Oliveira   University of Tras-os-Montes and Alto Douro,
                                   Portugal
Pedro Gonzalez   University of Jaen, Spain
Philip Samuel   Cochin University of Science and Technology,
                                   India
Pierre-Francois Marteau   Universite de Bretagne Sud, France
Pietro Ducange   University of Pisa, Italy
Punam Bedi   University of Delhi, India
Qieshi Zhang   Waseda University, Japan
Qinghan Xiao   Defence R&D Canada, Canada
Radu-Codrut David   Politehnica University of Timisoara, Romania
Rafael Bello   Universidad Central de Las Villas, Cuba
Ramin Halavati   Sharif University of Technology, Iran
Ramiro Barbosa   Instituto Superior de Engenharia do Porto,
                                   Portugal
Ramon Sagarna   University of Birmingham, UK
Richard Jensen   Aberystwyth University, UK
Robert Berwick   Massachusetts Institute of Technology, USA
Roberto Armenise   Poste Italiane, Italy
Robiah Yusof   Universiti Teknikal Malaysia Melaka, Malaysia
Roman Neruda   Institute of Computer Science, Czech Republic
S. Ramakrishnan   Dr. Mahalingam College of Engineering
                                   and Technology, India
Sabrina Ahmad   Universiti Teknikal Malaysia Melaka, Malaysia
Sadaaki Miyamoto   University of Tsukuba, Japan
Santi Llobet   Universitat Oberta de Catalunya, Spain
Sarwar kamal   East West University, Bangladesh
Satrya Fajri Pratama   Universiti Teknikal Malaysia Melaka, Malaysia
Saurav Karmakar   Georgia State University, USA
Sazalinsyah Razali   Universiti Teknikal Malaysia Melaka, Malaysia
Sebastian Ventura   University of Cordoba, Spain

Selva Rivera	Institute of C.E.D.I.A.C, Argentina
Shang-Ming Zhou	University of Wales Swansea, UK
Siby Abraham	University of Mumbai, India
Silvia Poles	EnginSoft, Italy
Silvio Bortoleto	Federal University of Rio de Janeiro, Brazil
Siti Rahayu Selamat	Universiti Teknikal Malaysia Melaka, Malaysia
Steven Guan	Xi'an Jiaotong-Liverpool University, China
Sung-Bae Cho	Yonsei University, Korea
Swati V. Chande	International School of Informatics and Management, India
Sylvain Piechowiak	Universite de Valenciennes et du Hainaut-Cambresis, France
Subhas Mukhopadhyay	Massey University, New Zealand
Takashi Hasuike	Osaka University, Japan
Taras Kotyk	Ivano-Frankivsk National Medical University, Ukraine
Tarek Gaber	Suez Canal University, Egypt
Tay Kai Meng	Universiti Malaysia Sarawak, Malaysia
Teresa Ludermir	Federal University of Pernambuco, Brazil
Thomas Hanne	University of Applied Sciences Northwestern Switzerland, Switzerland
Tzung-Pei Hong	National University of Kaohsiung, Taiwan
Ting-Ting Wu	National Yunlin University of Science and Technology, Taiwan
Vaclav Snasel	VSB-Technical University of Ostrava, Czech Republic
Valentina Colla	Scuola Superiore Sant'Anna, Italy
Victor Hugo Menendez Dominguez	Universidad Autonoma de Yucatan, Mexico
Vincenzo Loia	University of Salerno, Italy
Vincenzo Piuri	University of Milan, Italy
Virgilijus Sakalauskas	Vilnius University, Lithuania
Vivek Deshpande	MIT College of Engineering, India
Vladimir Filipovic	University of Belgrade, Serbia
Wahiba Ben Abdessalem Karaa	Taif University, KSA
Wei Wei	Xi'an University of Technology, China
Wei-Chiang Hong	Oriental Institute of Technology, Taiwan
Wen-Yang Lin	National University of Kaohsiung, Taiwan
Wilfried Elmenreich	University of Klagenfurt, Austria
Yasuo Kudo	Muroran Institute of Technology, Japan
Ying-Ping Chen	National Chiao Tung University, Taiwan
Yun-Huoy Choo	Universiti Teknikal Malaysia Melaka, Malaysia
Yunyi Yan	Xidian University, China
Yusuke Nojima	Osaka Prefecture University, Japan

Feng-Cheng Chang	Tamkang University, Taiwan
Yueh-Hong Chen	Far East University, Taiwan
Hsiang-Cheh Huang	National University of Kaohsiung, Taiwan
Yuh-Yih Lu	Minghsin University of Science and Technology, Taiwan

Sponsoring Institution

Fujian University of Technology, China
Fujian Provincial Key Laboratory of Big Data Mining and Applications, China

Contents

Intelligent Data Analysis
and Processing

The Complex Vector Maxwell Equations and an Applied Research

Miaoyu Zhang[1,2(✉)], Baolong Guo[1], and Jie Wu[2]

[1] Institute of Intelligent Control and Image Engineering,
Xidian University, Xian 710071, Shaanxi, China
myzhang1028@163.com
[2] School of Electronic Engineering, Xian Shiyou University,
Xian 710065, Shaanxi, China

Abstract. Different forms of Maxwell equations can clearly describe macroscopic electromagnetic laws of different problems. The complex vector Maxwell equations are deduced on the basis of the plural form equations. They visually show a process and a rule that a time-varying electromagnetic field is stimulated by a harmonic current source. Firstly, with reference to the complex vector Maxwell equations, the author analyzes basic rules and characteristics of the electromagnetic field that current source excites in the infinite conductive medium. It reveals an interdependent mechanism among the current, magnetic and electric field. Secondly, they are applied to the analysis of electromagnetic and current characteristics that a coil current source generates in induction logging around the borehole. The results show that the complex vector Maxwell equations not only clearly describe a physical relationship of mutual dependence and mutual excitation among the real vector and imaginary vector of the electric-field intensity, magnetic field intensity, induced current, displacement current and excitation current, but also deeply appears a relationship between the receiving voltage and the formation parameters in induction logging. The numerical calculation and drawing graphics display a law of the real vector and imaginary vector of the electric field intensity, magnetic field intensity, induced current, displacement current and excitation current.

Keywords: Complex vector · Maxwell equation · Electric field · Magnetic field · Induced current · Displacement current

1 Introduction

Maxwell equations are basic equations in the macroscopic electromagnetic phenomena and they reflect a law of variation of electromagnetic field. In the existing literatures, a lot of people made some transformations to the mathematics of Maxwell electrodynamics (Maxwell equations and Lagrange etc.) for various reasons and purposes, and interpreted their physical meaning differently. In recent years, a complex vector expression of electromagnetic field was proposed by Bing et al. [1]. They pointed out that electric field is a real part of a complex vector, magnetic field is an imaginary part, and a three-dimensional real vector in the traditional electromagnetism is rewritten as three-dimensional complex vector, thus the electric and magnetic fields are unified, the

J. Pan et al. (eds.), *Intelligent Data Analysis and Applications*, Advances in Intelligent Systems and Computing 535, DOI 10.1007/978-3-319-48499-0_1

complex vector equations of electromagnetic field are obtained. At the same time, A.I. Arbab also presented a unified complex model of Maxwell's equation, which resembles that of Xu Bing in research method. The form of Maxwell's equations is one vector equation and one scalar equation, which reveals the analogy existing between the quantum mechanical equations of motion [2]. In recent years, the author finds in the electromagnetic researching: When expressed in complex vectors, the plural form of complex amplitude vectors of electric field and magnetic field can clearly reveal how to excite and interconnect between the real part and the imaginary part of electric-field and magnetic field and establish a unified electromagnetic field. Then they can explained various mechanisms in induction logging. This article will derive the complex vector expression of Maxwell equations on the basis of this theory; the mutual relationship and influence factors between the electromagnetic quantities are analyzed in the establishment of the electromagnetic field and some conclusions are drawn.

2 Maxwell Equations

An alternating electric field and a magnetic field are not isolated. They are always closely linked together and excited each other, which makes a unified electromagnetic field. This is a basic concept of Maxwell electromagnetic theory in time-varying fields. In the sinusoidal electromagnetic field, the plural form of Maxwell equations are expressed as [3]

$$\nabla \times \dot{\vec{H}} = \dot{\vec{J}}_c + \dot{\vec{J}}_d + \dot{\vec{J}}_e. \tag{1}$$

$$\nabla \times \dot{\vec{E}} = -j\omega \dot{\vec{B}}. \tag{2}$$

$$\nabla \bullet \dot{\vec{B}} = 0. \tag{3}$$

$$\nabla \bullet \dot{\vec{D}} = \dot{\rho}. \tag{4}$$

The corresponding constitutive relations for linear electromagnetic media are expressed as

$$\dot{\vec{D}} = \varepsilon \dot{\vec{E}}. \tag{5}$$

$$\dot{\vec{B}} = \mu \dot{\vec{H}}. \tag{6}$$

$$\dot{\vec{J}}_c = \sigma \dot{\vec{E}}. \tag{7}$$

The time factor "$e^{j\omega t}$" is omitted from formula (1) to (7), all of the variables are plural forms. Among them, $\dot{\vec{E}}$ is the electric field intensity, the unit is V/m; $\dot{\vec{H}}$ is the magnetic field intensity; the unit is A/m; $\dot{\vec{B}}$ is the magnetic induction intensity, the unit is Wb/m^2; $\dot{\vec{D}}$ is the electric displacement vector, the unit is C/m^2; $\dot{\vec{J}}_c$ is the conducting current density, the unit is A/m^2; $\dot{\vec{J}}_d$ is the displacement current density, the unit is A/m^2, $\dot{\vec{J}}_d = j\omega\dot{\vec{D}}$; $\dot{\vec{J}}_e$ is the excitation current source, the unit is A/m^2; $\dot{\rho}$ is the charge density, the unit is C/m^3; ω is the angular frequency, the unit is rad/s; j is the imaginary unit. μ is the magnetic permeability, the unit is H/m; ε is the dielectric constant, the unit is F/m; σ is the electric conductivity, the unit is S/m.

In the loss medium that it is unbounded in space and filled with dielectric constant ε, magnetic permeability μ, electric conductivity σ, when there is passive $(\dot{\vec{J}}_e = 0, \dot{\rho} = 0)$, the electric field intensity $\dot{\vec{E}}$ of the plane electromagnetic wave is expressed as [4]

$$\dot{\vec{E}} = \dot{\vec{E}}_0 e^{-j\vec{k}\cdot\vec{r}}. \tag{8}$$

In the Eq. (8), $\dot{\vec{E}}_0$ is a complex amplitude vector of the electric field intensity $\dot{\vec{E}}$; $\vec{k}$ is the wave vector; $\vec{r}$ is the radial vector in the observation point.

It is customary to substitute k for $\gamma = jk$, γ is known as wave propagation constant, $\gamma = \alpha + j\beta$.

If electric wave spread along the z-direction and the initial phase is 0, the Eq. (8) turns into

$$\dot{\vec{E}} = \hat{x}E_0 e^{-\alpha z} e^{-j\beta z} = \hat{x}(E_0 e^{-\alpha z}\cos\beta z - jE_0 e^{-\alpha z}\sin\beta z). \tag{9}$$

Among them,

$$\alpha = \frac{\omega\sqrt{\mu\varepsilon}}{\sqrt{2}}\left\{\left[1+(\frac{\sigma}{\omega\varepsilon})^2\right]^{\frac{1}{2}}-1\right\}^{\frac{1}{2}}. \tag{10}$$

$$\beta = \frac{\omega\sqrt{\mu\varepsilon}}{\sqrt{2}}\left\{\left[1+(\frac{\sigma}{\omega\varepsilon})^2\right]^{\frac{1}{2}}+1\right\}^{\frac{1}{2}}. \tag{11}$$

We can see that the greater ω and σ are, the greater α and β are [4]. The amplitude of electric field intensity $\dot{\vec{E}}$ decays by exponent $e^{-\alpha z}$, the size is a plural, the vibration direction is x-direction, so it is called complex amplitude vector. α represents an attenuation constant, the unit is Np/m. β represents a phase shift constant, the unit is rad/m. Similarly, they have same characteristics on conduction current density $\dot{\vec{J}}_c$, displacement current density $\dot{\vec{J}}_d$, excitation current source $\dot{\vec{J}}_e$, magnetic field intensity $\dot{\vec{H}}$, magnetic induction intensity $\dot{\vec{B}}$ and electric displacement vector $\dot{\vec{D}}$.

3 Complex Vector Maxwell Equations

In Cartesian coordinates, an expression of the electric field intensity $\dot{\vec{E}}$ with plural form is

$$\dot{\vec{E}} = \hat{x}\dot{E}_x + \hat{y}\dot{E}_y + \hat{z}\dot{E}_z. \tag{12}$$

In the Eq. (12), $\dot{E}_x$, $\dot{E}_y$ and $\dot{E}_z$ are complex amplitudes in the direction x, y and z. The real and imaginary parts of the complex amplitude are brought into Eq. (12), the Eq. (13) is obtained.

$$\begin{aligned}
\dot{\vec{E}} &= \hat{x}(E_{xR} + jE_{xX}) + \hat{y}(E_{yR} + jE_{yX}) + \hat{z}(E_{zR} + jE_{zX}) \\
&= (\hat{x}E_{xR} + \hat{y}E_{yR} + \hat{z}E_{zR}) + j(\hat{x}E_{xX} + \hat{y}E_{yX} + \hat{z}E_{zX}) \\
&= \vec{E}_R + j\vec{E}_X.
\end{aligned} \tag{13}$$

Equation (13) is an electric-field complex vector in the harmonic fields. It shows that electric-field intensity can be expressed as complex vector which is composed of a real part and an imaginary vector. All of the plural form of variables can be written by complex vector in the same way [5]. The complex vector expression of variables are brought into Eqs. (1) and (2), the Eqs. (14) and (15) are obtained.

$$\nabla \times \vec{H}_R + j\nabla \times \vec{H}_X = \vec{J}_{cR} + \vec{J}_{dR} + \vec{J}_{eR} + j(\vec{J}_{cX} + \vec{J}_{dX} + \vec{J}_{eX}). \tag{14}$$

$$\nabla \times \vec{E}_R + j\nabla \times \vec{E}_X = \omega\mu\vec{H}_X - j\omega\mu\vec{H}_R. \tag{15}$$

The real and imaginary parts are separated in Eqs. (14) and (15), the induction current and displacement current are also written, the real form of the Maxwell equations are obtained.

$$\nabla \times \vec{H}_R = \vec{J}_{cR} + \vec{J}_{dR} + \vec{J}_{eR}, \vec{J}_{cR} = \sigma\vec{E}_R, \vec{J}_{dR} = -\omega\varepsilon\vec{E}_X. \tag{16}$$

$$\nabla \times \vec{H}_X = \vec{J}_{cX} + \vec{J}_{dX} + \vec{J}_{eX}, \vec{J}_{cX} = \sigma\vec{E}_X, \vec{J}_{dX} = \omega\varepsilon\vec{E}_R. \tag{17}$$

$$\nabla \times \vec{E}_R = -\omega\mu\vec{H}_X. \tag{18}$$

$$\nabla \times \vec{E}_X = -\omega\mu\vec{H}_R. \tag{19}$$

How to understand physical significance on Maxwell Eqs. (16)–(19) with real vector and imaginary vector of the complex vector?

The coil current generates sinusoidal electromagnetic field in the infinite conductive medium. The phase of the excitation current source is 0, $\dot{\vec{J}}_e = \vec{J}_{eR}$, We analyze a building process of electromagnetic field through Eqs. (16)–(19), as shown in Fig. 1.

1. According to Eq. (16), a real part of primary magnetic field is generated by the real part of current source, with the same phase as that of current source and the direction of the real part of primary magnetic field perpendicular to the current source.
2. According to Eq. (19), an imaginary part of primary electric field is generated by the real part of primary magnetic field, which direction perpendicular to the real part of magnetic field and opposite to current source.
3. According to Eq. (17), the imaginary part of the primary induction current is generated by the imaginary part of primary electric field, which direction opposite to that of the current source. It is proportional to conductivity. In the general conductive medium, $\frac{\sigma}{\omega\varepsilon} \gg 1$, the real part of displacement current is much smaller than the imaginary part of the induction current, so it can be ignored.
4. The imaginary part of the secondary magnetic field is generated by the imaginary part of the primary induction current, which direction opposite to the real part of primary magnetic field.
5. According to Eq. (18), the real part of the secondary electric field is generated by the imaginary part of the secondary magnetic field, which direction perpendicular to the imaginary part of magnetic field and opposite to the current source.
6. According to Eq. (16), the real part of the secondary induction current is generated by the real part of the secondary electric field.
7. The real part of the cubic magnetic field is generated by the real part of the secondary induction current, which direction opposite to the real part of the primary magnetic field, so the primary magnetic field is weakened.
8. According to Eq. (19), the imaginary part of the cubic electric field is generated by the real part of the cubic magnetic field, their direction are same.
9. The imaginary part of the cubic induction current is generated by the imaginary part of the cubic electric field, According to Eq. (17), the imaginary part of the quartic magnetic field is generated by the imaginary part of the cubic induction current, which direction opposite to the imaginary part of secondary magnetic field.
10. According to Eq. (18), the real part of the quartic electric field is generated by the imaginary part of the quartic magnetic field.
11. Repeat step (1).

The real part of current source excites the real part of primary magnetic field first. Then there are real part of the magnetic field, imaginary parts of the electric field and current in the odd field; there are imaginary part of the magnetic field, real parts of the

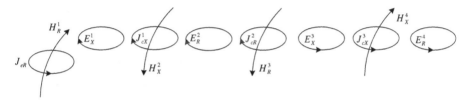

Fig. 1. The establishment of the electromagnetic field generated by the coil current source in an infinite conducting medium.

electric field and current in the even field. The magnetic field which radiate outward is excited by current source. Next the magnetic field which radiate inward is generated two times (the secondary and the cubic) by the current. It reflects and then it radiates outward. Power feedback phenomenon is explained in the loss medium by the field reflects inward [6]. Ohmic loss is generated by the induction current in the conductive medium [7]. So, electromagnetic fields and electromagnetic waves exist only within a certain range from the emission source because there are power feedback and Ohmic loss on the steady state.

4 Application in the Electromagnetic Induction Logging

The formation resistivity is measured by using electromagnetic induction principle in the electromagnetic induction logging areas. A transmitter passing a sinusoidal current in the borehole will generate a sinusoidal magnetic field, electric field and current that it is related to electromagnetic parameters of the formation medium. One or more receiving coils (array induction logging) are arranged from the transmitter coil in a certain distance. The information on medium are obtained through measuring received voltage. So, the formation characteristics are researched further and oil or gas is found. The receiving coil receives a complex voltage. It is often considered that the real part of the voltage is formation information, the imaginary part is the unwanted signal which doesn't include formation information, but its value is far greater than the real part. So, the imaginary signals are offset by a shielding coil winding opposite, otherwise they will drown the real signals [8]. The following electromagnetic induction logging mechanism is studied by the complex vector Maxwell equations and numerical calculation.

As shown in Fig. 2, it is assumed that normal directions on a transmitter and a receiver coil are z-direction in the homogeneous formation. The first phase of the current source is 0, $\dot{\vec{J}}_e = \vec{J}_{eR}$. It reveals an electromagnetic phenomenon around the formation what the complex vector Maxwell equations described above.

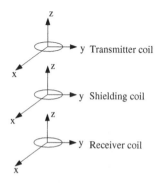

Fig. 2. Coils configuration schematic of the array induction tool.

The displacement current can be ignored when the inductions current is much larger than the displacement current $\left(\frac{\sigma}{\omega\varepsilon} \gg 1\right)$, ε approximately equal to ε_0, the operating frequency is low (such as the following will be analyzed at 13 kHz) and conductivity range is 0.0001 S/m–10 S/m in the induction logging. The following an establishment of magnetic field, electric field, conduction current is studied and a relationship between each field and formation conductivity according to complex vector Maxwell equations.

The real part of the magnetic field is comprised of odd magnetic field, and the imaginary part is even. The real part of the electric field is a superposition on even electric field, and the imaginary part is odd. Current is consistent with the electric field, but the size is not the same. The real and imaginary part of the received voltage corresponds to the real and imaginary part of electric field, and they correspond to the imaginary and real parts of magnetic field. The imaginary part of the secondary magnetic field and the real part of the secondary electric field are proportional to the formation conductivity. Their phases are different from emission current source at $-90°$ and $-180°$. The real part of the primary magnetic field and the imaginary part of the primary electric field have nothing to do with formation conductivity. Their phases are different from emission current source at $90°$ and $-180°$. But the real part of the magnetic field and the imaginary part of the electric field are high-order (cubic or quintal) and odd, which are related to the formation conductivity. Their intensity diminishes gradually with respect to the primary field. So, either the real part or the imaginary part of the voltage contains the formation information. Just the maximum signal only appears in the real part. The imaginary part of the voltage is complex because it is generated by the high-order electromagnetic field, unlike the real part of the voltage which is directly generated by the formation conductivity. It explains the reasons why influences on measuring the imaginary part of voltage are complex.

The distribution rules are analyzed that electric field, magnetic field, conduction current and displacement current are generated by a transmitting coil of the tri-axial array induction in z-direction around the center hole through the COMSOL software [9]. The model parameters are mainly as follows: borehole diameter is 0.2032 m (8 in.), instrument radius is 0.046 m, emission current is 1 A, frequency is 13 kHz, mud conductivity is 1.0 S/m in the borehole, formation conductivity is 0.01 S/m and instrument rod is filled with conductivity at 0 S/m because there is no mud in it. Figures 3, 4 and 5 illustrate variation characteristics of the real and the imaginary vector of three complex vectors at a section, including electric field, magnetic field and conduction current.

4.1 Electric Field Intensity $\vec{E}$

Figure 3 is normalized electric field line on the real vector $\vec{E}_R$ in section *xoy*. It is showed that a stable electric field is formed in the borehole and around the borehole; the electric field lines are rotationally symmetrical about the borehole center, the direction of the real electric field opposite to the excitation current source. It is consistent with Fig. 1.

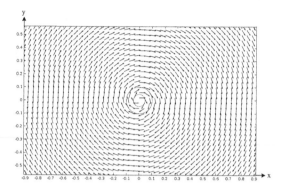

Fig. 3. Electric field line on the real vector $\vec{E}_R$ of the electric field intensity $\dot{\vec{E}}$ in section *xoy*.

4.2 Magnetic Field Intensity $\dot{\vec{H}}$

Figure 4 is normalized magnetic field line on the imaginary vector $\vec{H}_X$ of magnetic field intensity $\dot{\vec{H}}$ in section *yoz* when the uniform formation conductivity is 10.0 S/m. As we can see the imaginary vector of the magnetic field is generated by the conduction current distributing in the whole conductive formation, magnetic line of the imaginary is also distributed in the whole formation.

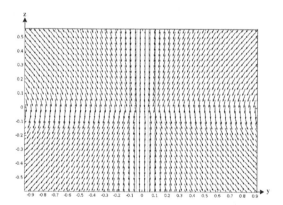

Fig. 4. Magnetic field line on the imaginary vector $\vec{H}_X$ of the magnetic field intensity $\dot{\vec{H}}$ in section *yoz*.

4.3 Conduction Current Density Vector $\dot{\vec{J}}_c$

Figure 5 is normalized electric field line on the real vector $\vec{J}_{cR}$ of conduction current density $\dot{\vec{J}}_c$ in section *xoy*. As we can see, a horizontal eddy is formed, which is in

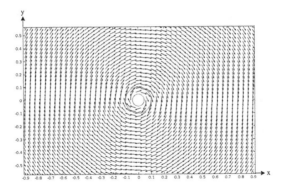

Fig. 5. Streamlines on the real vector $\vec{J}_{cR}$ of the induction current $\dot{J}_c$ in section *xoy*

accord with the direction of electric field and rotationally symmetrical about the instrument axis in the borehole and around the conducting medium [10], the conductivity is 0 in the instrument, and there is no induction current.

5   Conclusion

The electromagnetic models are established in the infinite conductive medium through complex vector Maxwell equations. The mechanism of electromagnetic induction logging is explained by the complex vector Maxwell equations, the following conclusions can be drawn:

1. The complex vector Maxwell equation clearly describes a physical relationship of mutual dependence and mutual excitation between the real part vector and imaginary vector on the complex amplitude vector of the electric field intensity, magnetic field intensity, induced current, the displacement current and excitation current.
2. The electromagnetic field rules what the coil current source generates are clearly explained through the complex vector Maxwell equation in the infinite conductive medium. The real part of the coil current source excites the real part of odd magnetic field, the imaginary parts of electric field and current, the imaginary part of even magnetic field, the real parts of electric field and current in the infinite conductive medium. The magnetic field which a coil current excites radiates outward from source. Then when the current excites the inward magnetic field every two times, the current radiates outward two times. It makes steady field only in certain range from the emission source because the Ohmic loss is generated by the reflection and induction current in the conductive medium.
3. The electromagnetic phenomena on transmitting coil of the tri-axial array induction logging tool in z-direction around the borehole are analyzed through the complex vector Maxwell equations. It is revealed a relationship between the receiving voltage and formation parameters. And streamlines of the real and the imaginary vector are drawn through numerical calculation. It visually demonstrates distribution characteristics of the electromagnetic field around the borehole.

References

1. Bing, X., Liang, Y., Lihua, L.: The complex vector expression of electromagnetic field. College Phys. **26**(4), 16–23 (2007)
2. Arbab, A.I.: Complex Maxwell's equations. Chin. Phys. B **22**(3), 030301-1-030301-6 (2013)
3. Kraus, F.: Electromagnetics with Applications, 5th edn, pp. 112–114. Tsinghua University Press, Beijing (2001)
4. Lidong, C., Jie, W., Zhongyi, W.: The Foundation of the Engineering Electromagnetic, pp. 23–39. Northwestern Polytechnical University Press, Xian (2002)
5. Mott, H., Dudgeon, J.E.: Complex solutions to Maxwell's equation. J. Frankl. Inst. **294**(1), 49–56 (1972)
6. Ymmamoto, Y., Yamaguchi, K.: Feedback effect for wireless high-power transmission. WSEAS Trans. Circ. Syst. **13**, 241–245 (2014)
7. Harmuth, H.F.: Propagation velocity of electromagnetic signals. IEEE Trans. Electromagn. Compat. **EMC-28**(4), 270–272 (1986)
8. Gengji, Z.: Electrical Logging, pp. 32–37. China University of Petroleum Press, Beijing (1996)
9. Zimmerman, W.B.J., CnTech Co., Ltd.: Modeling and Analysis of Multi-physics Field by COMSOL Multiphysics, pp. 52–85. China Communications Press, Beijing (2007)
10. Alotto, P., Gruosso, G., Moro, F.: Three-dimensional eddy current analysis in unbounded domains by a DEM-BEM formulation. COMPEL – Int. J. Comput. Math. Electr. Electron. Eng. **27**(2), 460–466 (2008)

Passenger Hailing Safety PASW Modeler and Big Data Statistical Analysis Study

S.H. Meng[1(✉)], A.C. Huang[2], T.J. Huang[3], J. Chen[1], and J.S. Pan[1]

[1] College of Information Science and Engineering,
Fujian University of Technology, Fuzhou, Fujian 350118, China
menghui@fjut.edu.cn
[2] Department of Electrical Engineering,
National Sun Yat-Sen University, Kaohsiung, Taiwan
[3] Guo-Guang Laboratory School,
National Sun Yat-Sen University, Kaohsiung, Taiwan

Abstract. This paper presents a study based on passenger hailing safety big data collection and PASW statistical analysis. A regression analysis model was used on data collected to study whether two or more variables were correlated. Changes in the direction and strength of correlation, and a regression analysis of arguments given estimates for the conditional expectation of the dependent variables, fully revealed the complex dependence. In addition, the RSS, which reflects the influence of random errors on dependent variables, measured the influence of the variance of factors other than passengers' hailing safety, data collection, and statistics analysis. In the linear regression analysis model, to improve traffic safety prediction and control, R2 represents the contribution rate of analytic variables to a forecast change.

Keywords: Big data · Regression analysis model · Residuals Sum of Squares (RSS or SSE) · Statistical product and · Service Solutions (PASW or SPSS)

1 Introduction

Traffic safety for drivers, passengers, and pedestrians has caused widespread concern. According to a big data analysis in a traffic report released by the Traffic Police in Nanjing, China, more than 40 % of the traffic accidents resulted from drivers failing to focus or concentrate on driving. This resulted in casualties, most of whom were youth. According to relevant statistics, the annual number of deaths in traffic accidents worldwide reached approximately 600,000, and up to 12,000,000 people were injured in car accidents. Therefore, casualties and financial losses caused by car accidents exceed that of fire, flood, and other disasters combined [1]. Car accidents are known as 'the no. 1 public hazard of the civilized world'.

According to the statistics, most car accidents occurred when drivers could not stay constantly focused during driving, were fatigued, or were hailed by passengers. Thus, the drivers could not concentrate on their driving, lane departures caused by careless small movements, and did not notice nearby vehicles. All of these caused safety implications for drivers, passengers, and pedestrians, and especially for public

© Springer International Publishing AG 2017
J. Pan et al. (eds.), *Intelligent Data Analysis and Applications*, Advances in Intelligent Systems and Computing 535, DOI 10.1007/978-3-319-48499-0_2

transportation drivers [2]. On the other hand, the study noted that many people hail cars anywhere along the road. Some even did so under dangerous conditions, such as in the driveways [3].

Based on a quantitative model analysis of passenger hailing safety, the safety of drivers, passengers, and pedestrians was studied by doing the following: taking safety as a dependent variable; using driving sight distance, hailing sight distance, and hailing behaviour as arguments; and building a linear regression model between dependent variables and arguments. To understand the influence of safety factors among them, the scope of research emphasized inductive analysis and surveys, intelligent data analysis, safety requirements for passengers and drivers, and core value mining. These data were obtained and converted into a professional value detection system [4].

2 Statistical Methods

To support a comprehensive passenger ride-hailing prompt system, we performed a large sample survey, date collection, storage, and data analysis, and curated the relevant data, is shown in Table 1 data statistics workflow.

Table 1. Data statistics workflow.

The first part deals with the basic information from the respondents, including gender, age range, and occupation. This part has multiple-choice questions to be answered by respondents.

The second part is a survey of the basic hailing behaviours of passengers, including hailing habits, attitudes towards hailing, and the demand for a smart driving assistance system. This part is to be answered by respondents based on their own personal riding experiences.

The third part is the basic driver behaviour survey, which includes driving habits, an understanding of the hailing methods of passengers, and their demands on a smart driving assistance system.

The fourth part takes the data collected through the design platform as impact factors, aggregates them into an information summary table, and establishes multiple variable linear regression analysis models using actual survey statistics [5–7]. Then, the survey results and complete data are analysed to obtain survey results with higher authenticity and reliability.

3   Data Entry: List of Samples Included in Data

(1) Respondent's gender: males account for 61.94 % of the total survey, females account for 38.06 %.

(2) Age distribution of respondents: in April 2014, the China Nanjing Traffic Police released a big-data analysis of traffic reports, of which the casualties were mainly youth. Based on relevant statistics, this study collected extensive samples in the range of 21–25 years old, 26–30 years old, and 31–40 years old as the study intervals.

(3) The driver has certain range of sight distance while driving. This has a certain impact on traffic safety. The driver not only has to constantly pay attention to the road conditions ahead, but also pay attention to whether there is a pedestrian ahead or if a passenger hails a ride. In addition, there are obstacles. Thus, the driving sight distance becomes very important. This is important for safety among drivers, passengers, and pedestrians. Therefore, statistics for driving sight distance are a very important consideration in traffic safety studies. According to statistics, 46.27 % of the crowd has a driving sight distance in sunny weather of up to 100 m, 47.76 % of the crowd has a sight distance up to 50 m, and only 5.9 % of people have a sight distance of approximately 20 m.

(4) Low visibility on rainy days, in addition to windshields covered with rain, are equivalent to the driver wearing a pair of dark sunglasses. Therefore, the statistics for driving sight distance on rainy days is very important to traffic safety studies. According to statistics, only 5.22 % of the crowd has a driving sight distance of up to 100 m on rainy days, 45.52 % of the crowd has a sight distance up to 50 m, and 49.25 % has a sight distance of approximately 20 m. For drivers who have to pay attention to traffic conditions and traffic safety, but are also distracted by looking for passengers hailing from the street, this increases the safety implications for drivers.

Passengers certainly expect to get a ride as soon as possible, and hope the drivers can drive them in shortest possible time. This can guarantee a car's loading rate, and meet the needs of prospective passengers. According to the statistical analysis, approximately 41.04 % of the crowd can notice people waving in the front, 13.43 % of the crowd does not notice people waving in the front, and the rest of the crowd (45.52 %) is to be determined depending on specific distance.

(5) On the other hand, in the statistics for passengers' hailing behaviours (by a survey of whether passengers hailed a taxi in the middle of the road, or at other unsafe locations), a total of 14.93 % of people often hailed a taxi at unsafe locations, 28.36 % would do it sometimes, while 19.04 % depended on the situation. Only 37.31 % of the people stated they would not have behaved in this manner under normal conditions.

Based on the statistics, the study considered hailing vehicles at a safe place as safe hailing behaviour. Depending on situations with unsafe hailing behaviour, we performed a probability estimation of China Taiwan, Mainland China, Japan, and the USA. This estimation uses the specific circumstances of population proportions, after being converted by probability. We obtained the information shown in the figure below. Statistics show that based on unsafe hailing behaviour estimated from population risk statistics, Mainland China's unsafe population exceeds nine billion. Thus, Mainland China is ranked first among the four countries listed.

4   Model Building and Analyses

Through the sample analysis, the study classified collected data into two categories (sight distances on sunny days and rainy days). Sample data (such as hailing vehicles at some unsafe locations, and so on) were included in a safety analysis of a passenger hailing prompt system, while sample data collected from hailing methods and hailing failures were classified in a passenger study using a multivariate linear regression mathematics model [5–7]. First, the study quantified the collected data and defined the variables. The study described the quantification of the variables with an example of the problems in the questionnaire. For example: Will you hail a taxi in the middle of road or at other unsafe places? A: Sometimes, B: Often, C: Never, D: Depends.

Then the study set the variables as follows, using 1, 2, 3, 4 instead of A, B, C, D as answers: 'Sometimes' as 1, 'Often' as 2, 'Never' as 3, and 'Depends' as 4. For the variables in multiple choice problems, the study used a multiple dichotomy method. The fundamental idea is to set each option in a problem as a variable, and separate each option into two options (select the option, or not select the option). For example, a multiple-choice question has three options A, B, and C. Select is 1, and unselect is 0. Using the quantification table information, the study could perform a Goodness of Fit Analysis, regression equation significancy test, and a regression coefficient significancy test to obtain the corresponding parameters. These were followed by a data analysis, summary, and conclusion.

5   PASW Safety Model Analysis

The dependent variable is safety. The variables are driving sight distance, hailing sight distance, and hailing behavior, is shown in Table 2.

Table 2. Variables[a] included/excluded.

Model	Variables included	Method
1	Haling behavior, Driving sight distance, Haling sight distance[b]	Input

a. Dependent variables: (safety)
b. All required variables included

Using Explained Sum of Squares define coefficient of multiple determination (CMD) in total square's ratio:

$$R^2 = \frac{U}{SST} \tag{1}$$

$R = \sqrt{R^2}$, it was called coefficient of multiple determination (CMD). The related relationships are more closely like Y and Independent variable $x^1,....x^m$, if R is more bigger. Ordinarily, it was considered to be related to the establishment of the relationship when R > 0.8.

Linear regression: the goodness of fit test is to determine the coefficient R2. The greater the value, the better the fit. By observing the adjusted coefficient of 0.840, the goodness of fit is high, is shown in Table 3. (The closer to 1, the more accurate the regression equation coefficients and parameters, and the better the fit for the regression. A goodness of fit above 0.8 is generally considered high.)

Table 3. Goodness of fit analysis.[b]

Model	R	R^2	Adjusted R^2	Estimated Standard Error
1	0.919[a]	0.844	0.840	1.542159864858501

a. Estimated variables: (constant)
b. Dependent variable: safety

The 'regression sum of squares' shows the explanatory parts of the variance of response variables by arguments contained in the regression model. The 'residual sum of squares' represents the variance of response variables that was not explained by the variables contained in the regression model. These two values are associated with the sample size and the number of arguments in the model. The larger the sample size, the greater the corresponding variance. df is the degree of freedom, which is the number of free value variables. F indicates the F test statistics, which are used to test the significance of the regression equation. Since the P value of the significance test of the regression equation is 0.000 (smaller than the significance level of 0.05), the linear relationship is pretty good, is shown in Tables 4 and 5. The significance test is passed. This tested the overall hypothesis and indicated that the sample data deduction of an actual population, and the overall null hypothesis, were significant and reasonable.

The points in the P-P diagram surround a line, which shows the residuals approximately obey a normal distribution. At this point, the regression model passed various tests and achieved a good fit. Therefore, a regression model (with safety as a dependent variable and driving sight distance, haling sight distance, and hailing behaviour as variables) to improve the safety of drivers and passengers was established

Table 4. Regression equation significance test.[a]

Model	Sum of Squares	dF	Mean Squares	F	Significance
Regression	1672.360	3	557.453	234.396	0.000
Residuals	309.173	130	2.378		
Sum	1981.534	133			

Table 5. Regression coefficients significance test.[a]

Model	Nonstandardized coefficients		Standard coefficient	t	Sig.
	B	Standard error			
(Constants)	1.333	0.495		2.691	0.008
Driving sight distance	0.110	0.005	0.785	22.060	0.000
Hailing sight distance	0.094	0.011	0.312	8.717	0.000
Hailing behavior	0.268	0.125	0.077	2.151	0.033

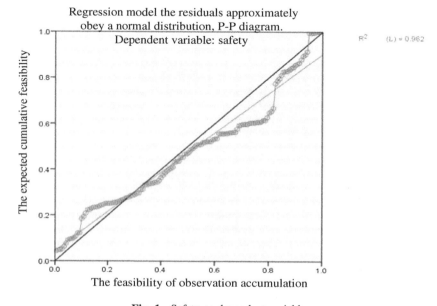

Fig. 1. Safety as dependent variable.

after analysis with $R^2 = 0.962$ shown in Fig. 1. This approaches 1 and with a diagnosis of residual normality.

Most of the points in the diagram gather along the line with a slope of 1 (ideal normal distribution line), which shows that the regression model passed the residual normal distribution test. This implies that the greater the overlap between the residual distribution curve and the normal distribution line, the higher the consistency between the two distributions. The established regression model has a positive significance.

6 Conclusions

Currently, no taxi has a prompt system that detects whether a passenger is hailing from the front. Instead, drivers are relied on to autonomously notice the presence of passengers who want a ride. Thus, drivers while driving must divide their attention to predict whether someone wants to hail the vehicle. This makes the drivers prone to traffic accidents, and can make drivers miss prospective passengers, thus reducing the loading rate.

This paper began with respondents' basic information, and used it to investigate people's driving and hailing behaviours, safety awareness, and demand analysis. Using statistical data from a population survey, a data analysis was used as basis to design hailing data. This was done for the safety of drivers and passengers, to obtain statistics that sensed distances, and to better ensure people's safety.

References

1. Yang, Z.: A Study of Measurement model for Economic Losses in Traffic Accidents. Shandong University of Science and Technology (2005)
2. Han, Y.: Pedestrian Detection System of the Advanced Driving Assistance Systems. Xidian University (2014)
3. Xu, H.: A Pedestrian Recognition Algorithm of a Visual Sensor Type in Vehicle Collision Prevention System. Guangdong University of Technology (2012)
4. Meng, S.H., Huang, A.C., Huang, T.J.: Passengers Hailing Reminding Method and Apparatus. Taiwan Patent (2016)
5. Tian, B.: Multiple linear regression analysis and its practical application. Yinshan Acad. J. (Natural Science Edition) (2011)
6. Cao, F.: Multivariate linear model parameter estimation. Wuhan University of Science and Technology (2006)
7. Wang, H., Meng, J.: Multiple linear regression analysis. J. Beijing Univ. Aeronaut. Astronaut. (2007)

Cellular Automaton Rule Extractor

Lukas Kroczek$^{(\boxtimes)}$ and Ivan Zelinka

Department of Computer Science, FEI, VSB Technical University of Ostrava,
Tr. 17. Listopadu 15, Ostrava, Czech Republic
{lukas.kroczek,ivan.zelinka}@vsb.cz
http://www.ivanzelinka.eu

Abstract. This article describes using an evolutionary algorithms to finding an original rule of generated 1D cellular automaton. In this article will be verified possibility and precision of searching rule from generated 1D cellular automaton using different evolutionary algorithms, random generators and rule definition. This extractor can be used for finding cellular automaton rule generating animals patterns like pigment patterns on the shells of mollusks.

Keywords: Cellular automaton · Evolutionary algorithms · SOMA · Differential evolution · Particle swarm optimization · Evolution strategies · Simulated annealing

1   Introduction

Cellular automaton (CA) are among the simplest mathematical representations of complex systems; where, for the moment, we may take complex system to mean any dynamical system that consists of more than a few typically non-linearly interacting parts. As such, CA are extremely useful idealizations of the dynamical behavior of many real systems, including physical fluids, neural networks, molecular dynamical systems, natural ecologies, military command and control networks, and the economy, among many others. Because of their underlying simplicity, CA are also powerful conceptual engines with which to study general pattern formation [1].

The simplest cellular automaton have a one dimensional array of cells with just two states have been studied by Wolfram [2].

The best-known way in which cellular automaton were introduced (and which eventually led to their name) was through work by John von Neumann in trying to develop an abstract model of self-reproduction in biology. Stephen Wolfram did extensive research in the properties of the simplest of cellular automaton; Elementary Cellular Automaton [3]. Every cell has two neighbors and two states. By a combinatorial approach of rules, there are 256 of such elementary cellular automaton.

Few elementary cellular automaton rules create similar structure to structure on the shell. Combining these cellular automaton rules with a little bit of noise will produce beautiful patterns with a bare minimum of computational effort.

© Springer International Publishing AG 2017
J. Pan et al. (eds.), *Intelligent Data Analysis and Applications*, Advances in Intelligent Systems and Computing 535, DOI 10.1007/978-3-319-48499-0_3

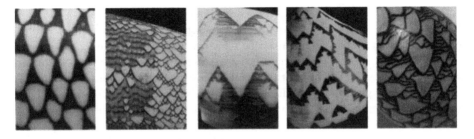

Fig. 1. Example of the pigment patterns on the shells of mollusks.

Using cellular automaton extractor introduced in this paper can be used to finding cellular automaton rule which generates pattern on the shells of mollusks or on other animals.

2   Experiment Design

The main idea of experiment and its results reported here is to verify possibility to find origin rule which generates 1D cellular automaton. Generated 1D cellular automaton will be clear (without a noise) to verify if idea is correct. In the experiment in this paper will be used 5 types of evolutionary algorithms, 3 types how we can specify rule and individual of used evolutionary algorithm, 2 types of random number generator. Experiments will use same configuration of cells in step 0.

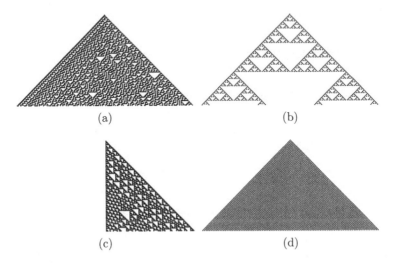

Fig. 2. Example of a generated cellular automaton. First 100 steps of rule 30 (a), rule 90 (b), rule 124 (c) and rule 250 (d).

The extractor is running selected evolutionary algorithm which an individual define the settings of the cellular automaton. The settings of the cellular automaton (for this paper) mean a rule which will be used for generating.

The algorithm starts with population of individuals, defined by the start world, that are evaluated by cost function. In the cost function of the evolutionary algorithm is generated a cellular automaton with a setting which is defined by individuals and count cost value of individuals as a sum of a differences between a values on the same position in the generated and the original cellular automaton. On the end of each iteration is choosed best individual (by minimal cost value) and when the evolutionary algorithm find the searched rule then searching is finished.

$$cost = \sum_{s=0}^{steps} \sum_{x=0}^{width} |v1_{s,x} - v2_{s,x}| \tag{1}$$

The experiment results reported in this paper use simplest two state cellular automaton. A world is generated for 10 thousands steps. Experiment run on virtual server with 6 virtual CPUs (Intel Xeon E5-2650 @ 2.6 GHz) and 4 GB RAM. Experiment application is written in C#.

Success in this paper mean the extractor algorithm find rule which is the origin rule of generated cellular automaton or a rule which can generates completely same cellular automaton for same settings. Time measures in tables define how long in maximum, minimum and average takes to the extractor finish searching of the origin rule.

In this paper we will test these configurations:

1. Evolutionary algorithm - this algorithm will be used for searching of origin rule.
   (a) Differential evolution (DE) [4]
   (b) Evolution strategies (ES) [5]
   (c) Simulated annealing (SA) [6]
   (d) Particle swarm optimization (PSO) [7]
   (e) Self-Organizing Migration Algorithm (SOMA) [8]
2. Rule/Individual definition - defines how will be rule of cellular automaton defined. Same definition will be used in individual. That mean how many dimensions and which values will individual contains.
   (a) Decimal
   (b) Binary
   (c) Gray code [9]
3. Random number generator - defines which type of random number generator will be used in evolutionary algorithms.
   (a) System - default random generator
   (b) Mersenne Twister [10]

3   Experiment

In the Table 1 are results for experiment focused on success rate of the evolutionary algorithms. The best evolutionary algorithms are DE and SOMA with almost same success rate. Focusing on search time we will find DE is more than 6.7 times faster than SOMA. In total DE is the fastest and SOMA is the slowest in average process time. Each evolutionary algorithm in the table run for 9216 times.

Table 1. This table shows results for each evolutionary algorithm ordered by success rate

Evolutionary alg.	Min [ms]	Max [ms]	Avg [ms]	Success [%]
DE	5116	453262	18420.15	99.91
SOMA	5226	853781	123815.59	99.90
PSO	5132	1826483	29739.34	99.61
SA	5148	4570131	122350.70	98.80
ES	6255	2284925	94560.50	98.06

In the Table 2 are results for experiment focused on the success rate of the rule (individual) definition type. The binary and the gray code don't have big difference in success rate. There are 15360 results for each definition.

Table 2. This table shows results for each rule/individual definition ordered by success rate

Rule/Ind. def.	Min [ms]	Max [ms]	Avg [ms]	Success [%]
Binary	5194	497208	49618.97	99.95
Gray code	5272	853781	63975.95	99.92
Decimal	5116	4570131	119736.85	97.90

In the Table 3 are results for experiment focused on the success rate by random generators. Configurations using the Mersenne twister was more succeeded but slower. We have 23040 runs of each random generator.

In the Table 4 are Top 10 results ordered by success rate. In the experiment are 3 configurations with 100 % success. All Top 10 configuration has the success rate bigger than 99.9 %. For all possible configurations is range of success rate was from 93.95 % to 100 %.

In the Table 5 is list of Top 10 results ordered by average time. All TOP10 fastest configurations use DE or PSO as evolutionary algorithm.

Table 3. This table shows results for each random generator ordered by success rate

Random generator	Min [ms]	Max [ms]	Avg [ms]	Success [%]
Mersenne	5116	4570131	80198.34	99.28
System	5132	4548868	75356.17	99.23

Table 4. This table shows Top 10 configurations ordered by success rate

Ev. alg.	Random	Rule/Ind.	Min [ms]	Max [ms]	Avg [ms]	Success [%]
PSO	System	Binary	5444	413045	14152.2	100
DE	System	Binary	5194	209073	18581.77	100
DE	Mersenne	Gray code	5288	193130	18652.98	100
ES	System	Binary	6271	373233	50482.46	100
SOMA	System	Binary	5444	497208	102680.44	100
SOMA	Mersenne	Binary	5397	357587	105165.19	100
SOMA	System	Gray code	5756	702101	110589.3	100
DE	Mersenne	Binary	5226	146672	17531.78	99.93
PSO	Mersenne	Gray code	5382	385948	18730.08	99.93
ES	System	Gray code	6458	285717	48737.43	99.93

Table 5. This table shows Top 10 configurations ordered by average time

Ev. alg.	Random	Rule/Ind.	Min [ms]	Max [ms]	Avg [ms]	Success [%]
PSO	Mersenne	Binary	5350	348741	14096.95	99.87
PSO	System	Binary	5444	413045	14152.2	100
DE	Mersenne	Binary	5226	146672	17531.78	99.93
DE	System	Decimal	5288	453262	18059.17	99.8
PSO	System	Gray code	5335	547519	18328.03	99.87
DE	Mersenne	Decimal	5116	192615	18492.31	99.87
DE	System	Binary	5194	209073	18581.77	100
DE	Mersenne	Gray code	5288	193130	18652.98	100
PSO	Mersenne	Gray code	5382	385948	18730.08	99.93
sDE	System	Gray code	5272	180306	19202.92	99.87

4 Summary and Conclusion

In this paper was verified possibility to find origin rule of generated cellular automaton. There was tested multiple configurations of the extractor and there was founded few configurations with 100 % success rate.

Our future work will be directed into improvement of process time and to verify possibility to find origin rule of generated cellular automaton with a noise.

Acknowledgment. The following grants are acknowledged for the financial support provided for this research: Grant of SGS No. SGS 2016/175, VSB-Technical University of Ostrava and by The Ministry of Education, Youth and Sports from the National Programme of Sustainability (NPU II) project "IT4Innovations excellence in science - LQ1602".

References

1. Ilachinski, A.: Cellular Automata: A Discrete Universe. World Scientific, London (2001)
2. Wolfram, S.: Universality and complexity in cellular automata. Physica **10D**, 1–35 (1984)
3. Wolfram, S.: Stephen Wolfram: A New Kind of Science. Wolfram Media, Champaign (2002)
4. Storn, R., Price, K.: Differential evolution - a simple and efficient heuristic for global optimization over continuous spaces. J. Global Optim. **11**, 341–359 (1997)
5. Beyer, H.G., Schwefel, H.P.: Evolution strategies - a comprehensive introduction. Nat. Comput. **1**, 3–52 (2002)
6. Kirkpatrick, S., Vecchi, M.P., et al.: Optimization by simmulated annealing. Science **220**, 671–680 (1983)
7. Kennedy, J.: Particle swarm optimization. In: Encyclopedia of Machine Learning, pp. 760–766. Springer, Heidelberg (2011)
8. Davendra, D., Zelinka, I.: Self-Organizing Migrating Algorithm. Springer, Heidelberg (2016)
9. Gray, F.: Pulse code communication. US Patent 2,632,058, March 1953
10. Matsumoto, M., Nishimura, T.: Mersenne twister: a 623-dimensionally equidistributed uniform pseudo-random number generator. ACM Trans. Model. Comput. Simul. **8**(1), 3–30 (1998)

Recovery of Compressed Sensing Microarray Using Sparse Random Matrices

Zhenhua Gan[1,2,4], Baoping Xiong[1,3], Fumin Zou[1,4(✉)],
Yueming Gao[3], and Min Du[2,3]

[1] College of Information Science and Engineering,
Fujian University of Technology, No. 3 Xueyuan Road, University Town,
Minhou, Fuzhou, Fujian, China
{ganzh, fmzou}@fjut.edu.cn, xiongbp@qq.com
[2] College of Electrical Engineering and Automation,
Fuzhou University, No. 2 Xueyuan Road, University Town,
Minhou, Fuzhou, Fujian Province, China
dm_dj90@163.com
[3] College of Physics and Information Engineering, Fuzhou University,
No. 2 Xueyuan Road, University Town, Minhou, Fuzhou, Fujian, China
fzugym@yahoo.com.cn
[4] Key Lab of Automotive Electronics and Electric Drive Technology
of Fujian Province, No. 3 Xueyuan Road, University Town,
Minhou, Fuzhou, Fujian, China

Abstract. Due to the uncertainty of elements in the random matrix, the design of composite probes on compressed sensing microarray (CSM) becomes more complexity. In this paper, we proposed a sparse random measurement matrix with '0/1' binary element, and fixed the same amount of elements '1' on each row, to construct the CSM composite probe. There is the same dilution for the mixed solution of target segments to ensure the consistency of gene concentration, so the composite probes which made up of the linear combination of target segments are very simple. Simulation experiment results show that the variation characteristics of the target segment can be accurately recovered by OMP algorithm under N = 96 sequence segments and variation sparsity level K ≤ 12, when M = 48 composite probes are constructed with a sparse random matrix fixed amount of non-zero elements each row.

Keywords: Compressed sensing · Sparse random matrix · Microarray · Composite probe · OMP

1 Introductions

The cDNA microarray has a large number of well defined sequence's probes which are integrated on the surface of the microarray's substrate. These probes will be hybridized simultaneously with the reference sequences and the test sequences, when these sequences are sufficiently amplified and labeled by fluorescent dye already. After cleaning and drying treatment, the intensity of the corresponding information will be

© Springer International Publishing AG 2017
J. Pan et al. (eds.), *Intelligent Data Analysis and Applications*, Advances in Intelligent Systems and Computing 535, DOI 10.1007/978-3-319-48499-0_4

read by the biological chip scanner. So we will have enough information to analyze the genetic sequences of biological tissues and cells [1].

In a traditional microarray, a large number of probe spots are arranged on the surface, but each probe represents a specific complementary gene segment which is used to detect the corresponding information [1]. In order to avoid the information losses caused by noise, multiple probes are usually used to carry out the repeated detections to improve the reliability. However, the multiple probes will increase the density of probe and reduced the size of spot, which also cause serious difficulties in obtaining reliable detection. Now, there is a more effective way to solve the above problems by designing the composite probe which could detect multiple gene segments simultaneously. And we will use the recovery algorithm to reconstruct the mutation information of each gene segment [2].

Compressed sensing (CS) is a novel sampling theory which is mainly composed of three parts, i.e., sparse signal, measurement matrix and reconstruction algorithm [3]. And signal or the signal via a specific transformation, with sparse or compressible characteristics, is the premise of the CS [4]. In order to ensure the accurate reconstruction of the signal easily, that is, without missing the valid information, the measurement matrix should be satisfied the restricted isometric property (RIP) [5].

Typical cDNA probes produce a large number of mostly useless information, due to the fact that the gene mutations of biological sequences are sparse. The compressed sensing microarray (CSM) is constructed based on the sparse gene mutation in the compressed sensing ideas. By literature [6], a method for constructing the compressed sensing composite probe was proposed. And the difference gene sequence could be recovered by the use of a small amount of the observation composite probes [7, 8].

2 Compressed Sensing and Sparse Random Matrices

2.1 Compressed Sensing

Compressed sensing theory points out that the signal x with K-sparse can be recovered by using a measurement matrix [9]. Consider a discrete digital signal $x \in R^N$ that can be expressed as $x = \psi^T f$, and f is K-sparse in the orthogonal basis $\psi \in R^{N \times N}$ We could design a measurement matrix $\emptyset \in R^{M \times N}$ while M << N. When the original signals are observed for M times, the observations will be shown as,

$$y_{M \times 1} = \phi_{M \times N} x = \phi_{M \times N} \psi_{N \times N} f_{N \times 1} = A_{M \times N} f_{N \times 1} \tag{1}$$

however, the Eq. (1) is an underdetermined system.

Due to the signal f is K-sparse and K < <N, Donoho, Candes, Tao et al., have already pointed out that we could obtain the accurate recovery of the signal f by minimizing the following type of l_0-norm.

$$\hat{f} = \arg \min \|f\|_0 \qquad s.t. \quad y = Af \tag{2}$$

where $\|f\|_0$ denotes the l_0-norm.

Since the solution of the l_0-norm is a combinatorial optimization problem, it is well known as a NP-hard problem. To avoid such computational difficulties, it is usually converted into an l_1-norm solution with the optimization constraints as long as the measurement matrix A satisfies the RIP. An l_1-norm solution is shown as following,

$$\hat{f} = \arg \min \|f\|_1 \qquad s.t. \quad y = Af \qquad (3)$$

Where $\|f\|_1 = \sum |f_i|$ with $i = 1,2,...,N$, is widely applied, and the formula (3) is usually converted into a linear programming problem.

2.2   Sparse Random Measurement Matrices

The compressed sensing measurement matrix is sparse for most applications, that is, the number of non-zero elements in the matrix is much smaller than the number of zero. Sparse random matrix has low complexity in encoding and reconstruction process. It can significantly reduce the computation and cause wide attention with the advantages of convenient updating and low storage capacity. Composite probes constructed by the sparse random matrix with 0/1 binary elements are very convenient. Especially, the measurement matrix fixed amount of element '1' in each row will make the composite probe combined of gene segments more simple. There is the same dilution for the mixed solution of probes to ensure the concentration consistency.

The sparse random M × N matrix contains $1 \leq l < N$ non-zero elements each row while $1 \leq t < M$ non-zero elements each column. Considering the signal x is K-sparse, and the literature [13] has provided that the sparse random matrix fixed amount of non-zero elements each row satisfies the RIP when the measurement $M \geq \min(\frac{NK}{lD^2}, \frac{N}{D^2})$ under '0/1' binary matrix element and the parameter D > 1. So the sparse random matrix fixed amount of element '1' each row is took the application for the CSM into account.

3   Compressed Microarray

3.1   Composite Probe of Compressed Microarray

The cDNA microarray uses the principle of molecular hybridization and gene complementation on the biologic gene probes. Since the reference samples and the test samples are already labeled by Cy3 and Cy5, they are specifically combining with the complementary biologic gene probes which are fixed on the microarray. We can get the fluorescent intensities via the biochip scanner, and then the gene segment's information of fluorescence intensities can be analyzed too. The cDNA microarray is usually labeled reference samples by Cy3, i.e. green marker, while the test samples are labeled by red dye of Cy5. The principle of traditional cDNA detection is shown as Fig. 1.

In a typical cDNA microarray, a large number of probe spots are located on the biochip surface, but each spot consists of a single gene sequence probe, which can only detect a specific complementary gene sequence. Due to the increasing density of

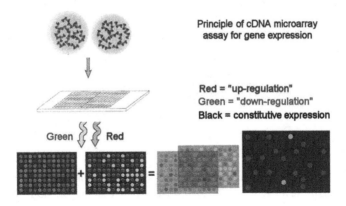

Fig. 1. The principle of traditional cDNA detection

microarray probe spots, which not only brings more difficulties at the manufacture of the gene spots, but also puts forward more stringent requirements to the biochip scanner. In contrast with the typical cDNA microarray, there is a more effective way to solve the above problems by designing the composite probe which could detect multiple gene segments simultaneously [2].

In order to simplify the construction of composite probes, we need to design a more efficient method to the CSM. In this way, we limit the elements of the random sparse matrix into binary 1/0 and fix amount of non-zero element in each row. The structure of the composite probe was illustrated in Fig. 2, and a compressed composite probe is represented as a linear combination of the each row of measurement matrix [6].

As shown in Fig. 2, a compressed composite probe is determined by the positions of the gene segments in the row of matrix, and the biochip scanner will read the intensity of fluorescence which is accumulated by a linear combination of the gene probes labeled by the fluorescent dye already. Considering N gene segments and $M \times N$ matrix A, we can easily construct M composite probes of CSM such as the k-th composite probe is determined by the k-th row of matrix A.

In two-color cDNA microarry, the reference sample is usually labeled by Cy3 while the test sample is labeled by Cy5 [14]. We are comparing two sample by vector x_{cy3}

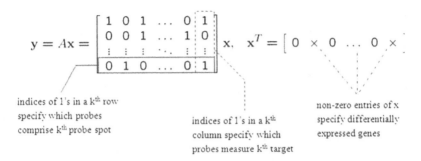

Fig. 2. Illustration of the compressed microarray

and vector x_{cy5}, and interesting in the difference expression of vector $x = x_{cy3} - x_{cy5}$. Since there are a few difference sequences in gene sequences, the vector x is K-sparse, $K \ll N$. In other words, the number of the difference sequences of gene segments with which we are concerned is smaller than the size of the vector x. We define a normalized observation value of the composite probe as vector $y = y_{cy3} - y_{cy5}$, so the combination structure of the composite probe is shown in the following,

$$y_k = \sum_{i=1}^{N} a_{ki} x_i, \; k = 1, \ldots, M, \quad \text{i.e.,} \quad y_{M \times 1} = A_{M \times N} x_{N \times 1} \qquad (4)$$

where $M \ll N$. The undermined equations of (4) may, in principle, be solved for the fact in the sparsity level $K \ll N$, so we could obtain x signal by the recovery of the formula (2) or (3).

3.2   Information Recovery of Composite Probes

Difference sequences recovery of composite probe, in essence, which is a reconstruction of compressed sensing with K-sparse characteristic. In the formula (2), it is theoretically optimal in terms of minimizing the number of measurements necessary for perfectly recovering any original signal x. However, directly carrying out the optimal solution in l_0-norm with a general measurement matrix A is NP-hard. In order to avoid such computational difficulties, we could choose an alternative approach, l_1-norm recovering shown on the formula (3), when the measurement matrix A satisfies the RIP. And further more, the classical sparse approximation methods to the l_1-norm recovering, such as the Orthogonal Matching Pursuit (OMP) algorithm, compared to the time-consuming convex optimization, would be very suitable [15].

The OMP algorithm

    Input   matrix A, measured value y, saprsity level K

    Initialize $\hat{x}_0 = 0, r_0 = y, j = 0, I = \Phi$

    Output   reconstruction of the signal $\hat{x}$

        (1) $j \leftarrow j + 1$;

        (2) Find the index that is the best match with residual vector r_{j-1},

            $\lambda_j \leftarrow \text{argmax}_i\{|< r_{j-1}, \; a_i >|\}$;

        (3) Update the index $I_j = I_{j-1} \cup \{\lambda_j\}$, and $A_j = [A_{j-1} \; a_j]$;

        (4) Reconstruction $\hat{x} \leftarrow [A_j]^{-1} y$;

        (5) Update the residual vector as $r_j \leftarrow y - A_j (\hat{x})$;

        (6) If $j \leq K$, then execute step (1), otherwise stop at $j > K$.

Fig. 3. Illustration of the OMP algorithm

In the OMP algorithm, considering $x_j = argmin_x \|y-A_j x\|_2$, $r_j = y-A_j x$, $A_j = [A_{j-1}\ a_j]$, which are selected in step j-th. Then, the OMP algorithm as shown on Fig. 3 [16, 17].

4   Simulation Results and Analysises

In this section, we present N = 96 simulation probes with the idea of cDNA, and the values of the reference probe vector x_{cy3} and the test probe vector x_{cy5} are subject to the random distribution individually. Furthermore, the difference segments vector $x = x_{cy3} - x_{cy5}$, also subject to random distribution.

We limit the vector x into the sparsity level K = 12. Figure 4 demonstrates the reference probe vector x_{cy3}, and Fig. 5 illustrates the probe vector x_{cy5}. Then, the differences between them, $x = x_{cy3}-x_{cy5}$ are shown in Fig. 6.

In the simulation experiments, we design a sparse random M × N matrix fixed amount of element '1' in each row as a compressed sensing measurement matrix A. We also make the arrangements of matrix A at N = 96, M = 48 and element sparse ratios $u = l/N = 0.25$, so there are M = 48 composite probes for the CSM that could be constructed by formula (4) from N = 96 gene segments.

Figure 7 illustrates the structure of the sparse random M × N matrix at N = 96, M = 48 and $u = 0.25$. And that, the normalized observation of the composite probes, y_{cy3} and y_{cy5}, are shown in Figs. 8 and 9, while the differences between them, $y = y_{cy3} - y_{cy5}$ are shown in Fig. 10.

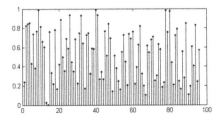

Fig. 4. The reference probe of x_{cy3}

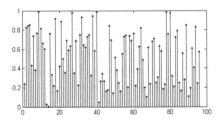

Fig. 5. The test probe of x_{cy5}

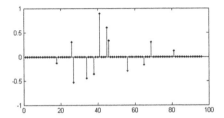

Fig. 6. The original differences signal $x = x_{cy3}-x_{cy5}$

Dark denoting '1' while white denoting '0'

Fig. 7. The sparse random 48 × 96 matrix A under $u = 0.25$

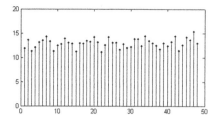

Fig. 8. The composite probes of y_{cy3}

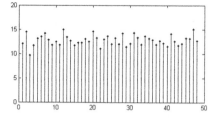

Fig. 9. The composite probes of y_{cy5}

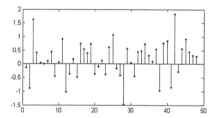

Fig. 10. The differences $y = y_{cy3}-y_{cy5}$

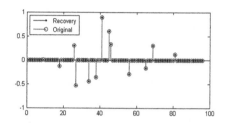

Fig. 11. The recovery of $\hat{x} = x_{cy3}-x_{cy5}$ with $e = 1.93 \times 10^{-15}$

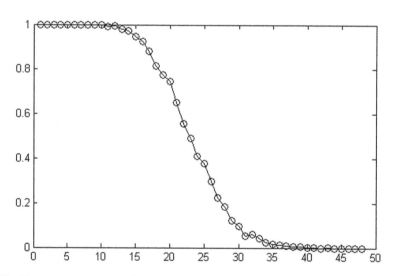

Fig. 12. The accurate reconstruction ratios of $\hat{x} = x_{cy3}-x_{cy5}$ at N = 96, M = 48 and u = 0.625 under $e \leq 1 \times 10^{-3}$

We use the OMP algorithm to accurately reconstruct the gene difference vector x, when N = 96, M = 48, K = 24, and u = 0.25. The reconstruction results with the relative error $e = 1.93 \times 10^{-15}$ are shown in Fig. 11.

We also consider that the parameters used in the simulations are N = 96, M = 48 and u = 0.625 while the alteration of K-sparse. The vectors $^\wedge x = x_{cy3}\text{-}x_{cy5}$ of the probes are recovered via OMP algorithm with the K varying from 0 to N, as shown in Fig. 12, which is indicated the reconstruction ratios on considering the accuracy, i.e., the relative error, $e \leq 1 \times 10^{-3}$.

5 Summary and Conclusions

Due to the sparse variation characteristics of gene sequence, the compressed sensing idea was adopted to the gene mutation in the composite probes of CSM. This idea not only reduces the probes density, but also easily realizes the multiple measurements for gene segments in a limited region of the microarry. For the uncertainty of the non-zero elements of the random matrix, the design of composite probes becomes more difficult. So we propose the sparse random measurement matrix fixed amount of element '1' in each row, which makes the composite probe combined of gene segments very simple. There is the same dilution for the mixed solution of probes to ensure the consistency of gene concentration. Simulation experiment results show that the variation characteristics of the sequence segments can be accurately recovered by OMP algorithm for compressed sensing under the variation quantity K $\leq$ 12, when M = 48 composite probes of CSM are constructed from N = 96 target segments by using a sparse random matrix fixed amount of element '1' each row.

Acknowledgments. This work is partially supported by the National Natural Foundation Project (61304199), the Ministry of Science and Technology projects for TaiWan, HongKong and Maco (2012DFM30040), the Major projects in Fujian Province (2013HZ0002-1,2013 YZ0002,2014YZ0001), the Science and Technology project in Fujian Province Education Department (JB13140/GY-Z13088), and the Scientific Fund project in Fujian University of Technology (GY-Z13005,GY-Z13125).

References

1. Guoliang, H., Chen, D., Shukuanl, X., et al.: Novel detection system of microbe chip and its application. Acta Optica Sinica. **27**(3), 499–504 (2007)
2. Shmulevich, I., Astola, J., Cogdell, D., et al.: Data extraction from composite oligonucleotide microarrays. Nucleic Acids Res. **31**(7), 431–439 (2003)
3. Wenze, S., Zhihui, W.: Advances and perspectives on compressed sensing theory. J. Image Graph. **17**(1), 1–12 (2012)
4. Jing-Wen, W., Xu, W.: Image reconstruction method based on compressed sensing for magnetic induction tomography. J. Northeast. Univ. (Nat. Sci.) **12**, 1687–1690 (2015)
5. Shi, G.M., Liu, D.H., Gao, D.H., Liu, Z., Lin, J., Wang, L.J.: Advances in theory and application of compressed sensing. Acta Electronica Sinica **37**(5), 1070–1081 (2009)
6. Parvaresh, F., Vikalo, H., Misra, S., et al.: Recovering sparse signals using sparse measurement matrices in compressed DNA microarrays. IEEE J. Sel. Topics Sig. Process. **2**(3), 275–285 (2008)

7. Sheikh, M.A., Sarvotham, S., Milenkovic, O., et al.: DNA array decoding from nonlinear measurements by belief propagation. In: 2007 IEEE/SP Workshop on Statistical Signal Processing, SSP 2007, pp. 215–219. IEEE (2007)
8. Parvaresh, F., Vikalo, H., Misra, S., Hassibi, B.: Recovering sparse signals using sparse measurement matrices in compressed DNA microarrays. IEEE J. Sel. Top. Sig. Process. **2**(3), 275–285 (2008)
9. Qiong-Hai, D., Chang-Jun, F.U., Xiang-Yang, J.I.: Research on compressed sensing. Chin. J. Comput. **34**(3), 425–434 (2011)
10. Gilbert, A., Indyk, P.: Sparse recovery using sparse matrices. Proc. IEEE **98**(6), 937–947 (2008)
11. Li, X.: Research on measurement matrix based on compressed sensing, pp. 16–19. Beijing Jiaotong University (2010)
12. Bo, Z., Yu-lin, L., Kai, W.: Restricted isometry property analysis for sparse random matrices. J. Electron. Inf. Technol. **1**, 169–174 (2014)
13. Jing-ming, S., Shu, W., Yan, D.: Lower bounds on the number of measurements of sparse random matrices. Sig. Process. **28**(8), 1156–1163 (2012)
14. Yang, X.U., Qiong-Fang, R., Yan-Ping, L.I.: Analysis methods of expression genes. J. Food Sci. Biotechnol. **27**(1), 122–126 (2008)
15. Tropp, J.A., Gilbert, A.C.: Signal recovery from random measurements via orthogonal matching pursuit. IEEE Trans. Inf. Theory **53**(12), 4655–4666 (2008)
16. Wang, J.: Support recovery with orthogonal matching pursuit in the presence of noise. IEEE Trans. Signal Process. **63**(21), 5868–5877 (2015)
17. Li, F., Guo, Y.: Compressed Sensing Analysis, pp. 66–69. Science Press, Beijing (2015)

The Reliability and Economic Analysis Comparison Between Parallel System and Erlang Distribution System

Lin Xu[1($\boxtimes$)], Chao-Fan Xie[2], and Lu-Xiong Xu[3]

[1] Economic Institute, Fujian Normal University, Fuqing, Fuzhou, Fujian, China
xulin@fjnu.edu.com
[2] Network Center, Fuqing Branch of Fujian Normal University,
Fuqing, Fuzhou, Fujian, China
119396356@qq.com
[3] Electronic Information and Engineering Institute,
Fuqing Branch of Fujian Normal University, Fuqing, Fuzhou, Fujian, China
xlx123456@139.com

Abstract. When people construct complex system, in order to improve the reliability of the system, they tend to take redundant backup or to take parallel the component to achieve it, but taking redundant backup, it needs to increase the system detection equipment and switch parts is necessary, which is a significant increase in the cost of the system, in the other hand, when we taking parallel system, it will reduce the cost, but also reduce the reliability of the system. This will compare the Exponential distribution of the parallel system reliability and redundancy systems, and analyzes increase the number of components or in parallel will bring about improvement of System Marginal reliability. To analyze In the economic cost constraints implement in parallel or redundant, and the impact of the economic cost of the switch to system, lastly it gives the best choice to choose whether redundant backup or parallel in some kind of condition and environment.

Keywords: Reliability · Redundancy · Exponential distribution · Parallel system

1 Introduction

Reliability theory is a discipline that can analyze characterize the products specified functions probability of occurrence of random events, which Is the sixties of last century developed new interdisciplinary subject. Thus, the reliability theory is based on probability theory, the first based on field research is machine maintenance problem [1]. At present, the main study of the reliability is the system reliability indices, as well as the reliability of the index on the basis of the optimal detection time to avoid faults, reduce the losses caused by the fault, such as the literature [2–4]. In N element exponentially distributed parallel system, Xie Chaofan and Xu Luxiong has been analyzed reliability and extreme value, obtained some relevant theoretical guidance.

© Springer International Publishing AG 2017
J. Pan et al. (eds.), *Intelligent Data Analysis and Applications*, Advances in Intelligent Systems and Computing 535, DOI 10.1007/978-3-319-48499-0_5

Without considering the economic constraints, the failure rate is equal to all of the components, the system reliability reaches a minimum, in considering the economic constraints, and when the failure rate is unit elastic, the conclusion is the same. And if you choose a good product in parallel with a poor product, the product reliability are more higher than parallel on two moderate of the failure rate. During operation, according to the actual situation, if the failure rate in the envelope line, then the whole system components should be replaced altogether, but if the failure rate is far away from the envelope, then for economic performance considerations, just to update the highest failure rate of several originals out [5]. This article will compares reliability of a parallel system which subject to the exponential distribution with redundancy system, analysis increase the number of elements or the number of parallel system brought the increase of marginal reliability, thereby selecting the economic cost of the constraints are implemented in parallel or redundant, and economic cost of the switch bring what the influences to the decision.

1.1   The Definition of the Main Indicators of Reliability

The contact author is asked to check through the final pdf to make sure that no errors have crept in during the transfer or preparation of the files. This should not be seen as an opportunity to update or copyedit the papers, which is not possible due to time constraints. Only errors introduced during the preparation of the files will be corrected.

(1) Reliability

The definition of reliability $R(t)$ [5]: it is the probability that product completes the required function under the specified conditions and within the prescribed time.

If the life distribution of product is $F(t)$, $t > 0$, the reliability $R(t) = P(T \geq t) = 1 - F(t)$. This is a function of time(t), so it can be called as reliability function. To the components obeying exponential distribution λ, its reliability is $e^{-\lambda t}, t \geq 0$.

(2) System parameter specification

A: represents normal working events of system.

A_i: represents normal working events of the element i.

λ_i: represents failure rate of the element i.

$R_s(t)$: represents system reliability, that is, $P(A) = R_s$.

$R_i(t)$: represents reliability of the element i, that is, $P(A_i) = R_i$.

m_s: represents the average lifetime of the system, $m_s = \int_0^{+\infty} R_s(t)dt$.

Parallel system: It is a system consisting of n components. As long as one of these elements works, the system can work; only when all the units fail, the system would fail.

According to the property of probability, the normal working probability of system $P(A) = P(\bigcup_{i=1}^{n} A_i)$ is as follows:

$$R_s(t) = 1 - \prod_{i=1}^{n} (1 - R_i(t)) = 1 - \prod_{i=1}^{n} (1 - e^{-\lambda_i t}) \tag{1}$$

Redundant backup systems, some of the elements work, the other unit does not work, in a waiting or a standby state, When the unit the failure rate of the secondary unit in a standby period is zero, in other words, is a hundred percent reliability during standby. One work, $n - 1$ standby redundant system framework is shown below Fig. 1, where, K is detected and switches.

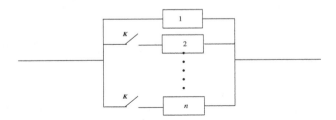

Fig. 1. The logical block diagram of Erlang distribution system.

The following lemma is assumed that when the switch is absolutely reliable:

Lemma 1.1: $X_1, X_2, \cdots X_n$ is mutually independent random variables, subject to the same parameter λ exponential distribution, then $X = X_1 + X_2 + \cdots + X_n$ obey Erlang distribution of order n, the probability density function is:

$$b_n(u) = \begin{cases} \lambda e^{-\lambda u} \frac{(\lambda u)^{n-1}}{(n-1)!}, & u \geq 0 \\ 0, & u < 0 \end{cases} \tag{2}$$

Proof: Let $p(u)$ is a probability density function of random variables, Since X_i are independent and identically distributed random variables exponentially distributed, then $p(u)$ is exponential distribution probability density function, to make $\varphi(t) = \int_{-\infty}^{+\infty} e^{itu} p(u) du$ is a characteristic function of random variable X_i then $\varphi(t) = (1 - \frac{it}{\lambda})^{-1}$. Because of $X = X_1 + X_2 + \cdots + X_n$, then X shows the probability density function of X_i do the n-fold convolution, Since the Fourier transform can become convolution to multiplication. Therefore, the characteristic function is:

$$\varphi_n(t) = \varphi(t)^n = (1 - \frac{it}{\lambda})^{-n}$$

The probability density function of X is, when $u \geq 0$, $b_n(u) = \frac{1}{2\pi} \int_{-\infty}^{+\infty} e^{-itu} \varphi_n(t) dt = \frac{1}{2\pi} \int_{-\infty}^{+\infty} e^{-itu} (1 - \frac{it}{\lambda})^{-n} dt$, points can be obtained from the Division:

$$\frac{1}{2\pi} \int_{-\infty}^{+\infty} e^{-itu} (1 - \frac{it}{\lambda})^{-n} dt = \frac{1}{2\pi} [\frac{\lambda}{(n-1)i} e^{itu} (1 - \frac{it}{\lambda})^{-n+1} \Big|_{-\infty}^{+\infty} + \frac{\lambda u}{n-1} \int_{-\infty}^{+\infty} e^{-itu} (1 - \frac{it}{\lambda})^{-n+1} dt]$$

$$= \frac{1}{2\pi} \frac{\lambda u}{n-1} \int_{-\infty}^{+\infty} e^{-itu} (1 - \frac{it}{\lambda})^{-n+1} dt = \cdots = \frac{1}{2\pi} \frac{(\lambda u)^{n-1}}{(n-1)!} \int_{-\infty}^{+\infty} e^{-itu} (1 - \frac{it}{\lambda})^{-1} dt = \lambda e^{-\lambda u} \frac{(\lambda u)^{n-1}}{(n-1)!}$$

So for the $n - 1$ redundancy elements and system of component life exponentially distributed, its distribution follows n-order Erlang distribution, proof.

Lemma 1.2: When the system life obey the n-order Erlang distribution, the reliability of system is

$$R_s(t) = [1 + \lambda t + \frac{(\lambda t)^2}{2!} + \cdots + \frac{(\lambda t)^{n-1}}{(n-1)!}]e^{-\lambda t} \tag{3}$$

Proof: By Lemma 1.1 shows that probability density function of the system life X is:

$$b_n(u) = \begin{cases} \lambda e^{-\lambda u} \frac{(\lambda u)^{n-1}}{(n-1)!}, & u \geq 0 \\ 0, & u < 0 \end{cases}$$

Since $R_s(t) = P(T \geq t) = 1 - F(t) = 1 - \int_0^t \lambda e^{-\lambda u} \frac{(\lambda u)^{n-1}}{(n-1)!} du$, points can be obtained from the Division:

$$\int_0^t \lambda e^{-\lambda u} \frac{(\lambda u)^{n-1}}{(n-1)!} du = -\int_0^t \frac{(\lambda u)^{n-1}}{(n-1)!} de^{-\lambda u} = -e^{-\lambda u} \frac{(\lambda u)^{n-1}}{(n-1)!} \Big|_0^t + \int_0^t \lambda e^{-\lambda u} \frac{(\lambda u)^{n-2}}{(n-2)!} du$$

$$= -e^{-\lambda t} \frac{(\lambda t)^{n-1}}{(n-1)!} + \int_0^t \lambda e^{-\lambda u} \frac{(\lambda u)^{n-2}}{(n-2)!} du = \cdots = [1 + \lambda t + \frac{(\lambda t)^2}{2!} + \cdots + \frac{(\lambda t)^{n-1}}{(n-1)!}]e^{-\lambda t} + 1$$

we can get the conclusion, proof.

2 Comparative Analysis of Reliability Distribution of Erlang and the Parallel System

Theorem 2.1: For the parallel system exponential distribution and Erlang distribution system, in parallel with Erlang order equal to the number of time, the average life expectancy differences between their two systems in parallel with the increase in the number increases, increasing equation is the following

$$\frac{n}{\lambda} - \frac{1}{\lambda}(1 + \frac{1}{2} + \frac{1}{3} \cdots + \frac{1}{n}) \tag{4}$$

Proof: $R_S^p(t), R_S^I(t)$ respectively corresponding parallel system and Erlang system reliability, then:

$$R_S^p(t) = 1 - (1 - e^{-\lambda t})^n, R_S^I(t) = [1 + \lambda t + \frac{(\lambda t)^2}{2!} + \cdots + \frac{(\lambda t)^{n-1}}{(n-1)!}]e^{-\lambda t}$$

$$\Delta s_n = \int_0^{+\infty} R_S^I(t) - R_S^p(t)dt = \int_0^{+\infty} \sum_{k=1}^n \frac{(\lambda t)^{k-1}}{(k-1)!} e^{-\lambda t}dt - \int_0^{+\infty} 1 - (1 - e^{-\lambda t})^n dt$$

since   the   gamma   function $\Gamma(s) = \int_0^{+\infty} x^{s-1}e^{-x}dx,$ then

$$\int_0^{+\infty} \sum_{k=1}^n \frac{(\lambda t)^{k-1}}{(k-1)!} e^{-\lambda t}dt = \sum_{k=1}^n \int_0^{+\infty} \frac{(\lambda t)^{k-1}}{(k-1)!} e^{-\lambda t}dt = \sum_{k=1}^n \int_0^{+\infty} \frac{(x)^{k-1}}{(k-1)!} e^{-x}dx$$

$$= \sum_{k=1}^n \frac{1}{\lambda(k-1)!} \int_0^{+\infty} x^{k-1}e^{-x}dx = \sum_{k=1}^n \frac{\Gamma(k)}{\lambda(k-1)!} = \frac{n}{\lambda}$$

$$\int_0^{+\infty} 1 - (1 - e^{-\lambda t})^n dt = \int_0^{+\infty} (1 - (1 - e^{-\lambda t}))(1 + (1 - e^{-\lambda t}) + \cdots (1 - e^{-\lambda t})^{n-1})dt$$

$$= \frac{1}{\lambda} \int_0^1 (1 + (1 - x) + \cdots (1 - x)^{n-1})dx = \frac{1}{\lambda}(1 + \frac{1}{2} + \frac{1}{3} \cdots + \frac{1}{n})$$

Therefore $\Delta s_n = \frac{n}{\lambda} - \frac{1}{\lambda}(1 + \frac{1}{2} + \frac{1}{3} \cdots + \frac{1}{n})$, proof.

Now using marginal analysis method analyze the circumstances of Δs_n with changing of n, $\Delta s_{n+1} - \Delta s_n = \frac{1}{\lambda}(1 - \frac{1}{n+1})$, this shows that Δs_n is the monotonically increasing sequence of n, but Δs_n is not the Cauchy column, it does not converge, then $\lim_{n\to\infty} \Delta s_n = +\infty$ so then we must consider the order of growth of Δs_n, there are theorems follows:

Theorem 2.2: For the parallel system of potential distribution and Erlang distribution system, the number of parallel and equal Erlang order, the average life expectancy differences between their two systems Δs_n the growing order is $O(n)$.

Proof: From Theorem 2.2, $\frac{n}{\lambda} - \frac{1}{\lambda}(1 + \frac{1}{2} + \frac{1}{3} \cdots + \frac{1}{n})$, from Euler's theorem, $1 + \frac{1}{2} + \frac{1}{3} \cdots + \frac{1}{n} = C + Inn + \varepsilon_n$, $C = 0.577216 \cdots$ called Euler's constant formulas, and when $n \to \infty$, $\varepsilon_n \to 0$, therefore ε_n is infinitesimal. Thus, Δs_n can be rewritten into the following equation:

$$\Delta s_n = \frac{1}{\lambda}(n - Inn - C + \varepsilon_n)$$

$$\lim_{n\to\infty} \frac{\Delta s_n}{n} = \lim_{n\to\infty} \frac{(n - Inn - C + \varepsilon_n)}{\lambda n} = \frac{1}{\lambda}$$

Therefore, the order of Δs_n is $O(n)$, since $\Delta s_{n+1} - \Delta s_n = \frac{1}{\lambda}(1 - \frac{1}{n+1})$, apparent differences in the average life expectancy of its two systems Δs_n, the rate of change is $\Delta^2 s_n = \frac{1}{n(n+1)}$, which order is $o(\frac{1}{n^2})$.

3 Reliability Economic Comparison Analysis of Parallel System and Erlang Distribution

Now we use the income from the system of a life cycle operation as an economic indicator. The Erlang system in the life cycle of the average yield for the L^I. Then

$$L^I = \frac{n}{\lambda}s - nc - (n-1)k \tag{5}$$

For the parallel systems, The average income of the system is:

$$L^P = \frac{s}{\lambda}(1 + \frac{1}{2} + \frac{1}{3}\cdots + \frac{1}{n}) - nc \tag{6}$$

s is unite time income, c is unite element cost, k is unite switch cost, owing to the L^I and L^P all contain the cost of element. Let λ as a parameter. In fact, Under the constraints of economic costs, Is to invest in the founds into a redundant backup or parallel, or to invest in the funds to choose a lower λ element all need the decision maker have to consider. Now consider the following model:

$$\begin{array}{ll} \max L^I = \frac{n}{\lambda}s - nc - (n-1)k & \max L^P = \frac{s}{\lambda}(1 + \frac{1}{2} + \frac{1}{3}\cdots + \frac{1}{n}) - nc \\ \text{s.t.} \, nf(\lambda) = u & \text{s.t.} \, nf(\lambda) = u \end{array} \tag{7}$$

$f(\lambda)$ is the failure rate and cost related functions, it is convex function and $f'(\lambda) < 0$ [6]. From constraint conditions $\lambda = f^{-1}(\frac{u}{n})$.

Then we can substitution the objective function and get that:

$$\begin{aligned} L^I &= \frac{n}{f^{-1}(\frac{u}{n})}s - nc - (n-1)k \\ L^P &= \frac{s}{f^{-1}(\frac{u}{n})}(1 + \frac{1}{2} + \frac{1}{3}\cdots + \frac{1}{n}) - nc \end{aligned} \tag{8}$$

And $L^I - L^P = \frac{n}{f^{-1}(\frac{u}{n})}s - (n-1)k - \frac{s}{f^{-1}(\frac{u}{n})}(1 + \frac{1}{2} + \frac{1}{3}\cdots + \frac{1}{n})$, Now divided $L^I - L^P$ into two parts: $L^I - L^P = (\frac{s}{f^{-1}(\frac{u}{n})} - k)(n-1) - \frac{s}{f^{-1}(\frac{u}{n})}(\frac{1}{2} + \frac{1}{3}\cdots + \frac{1}{n})$, $f'(\lambda) < 0$ and f^{-1} is monotonically decreasing too. With $n \uparrow$, $\frac{s}{f^{-1}(\frac{u}{n})} - k \downarrow$. So when the n is high enough, $L^I - L^P < 0$. The size of the turning point n_0 is related to the change of the curve $f^{-1}(\frac{u}{n})$. Because due to the Taylor formula we can know $\frac{1}{f^{-1}(\frac{u}{n})} = \left(\frac{1}{f^{-1}(\frac{u}{n})}\right)'\frac{u}{n} + o(\frac{1}{n})$, While to make n in a relatively large, $\left(\frac{1}{f^{-1}(\frac{u}{n})}\right)'\frac{u}{n} = k + \varepsilon$, ε is used to offset the effect of $\frac{s}{f^{-1}(\frac{u}{n})}(\frac{1}{2} + \frac{1}{3}\cdots + \frac{1}{n})$, and we will know that should be let cost u large enough, the detection switch k need more small, and $\frac{1}{f^{-1}(x)}$ near the 0 point, the change is very large, then we get $n \approx \left(\frac{1}{f^{-1}(\frac{u}{n})}\right)'\frac{u}{k}$. Otherwise, n can only take a small relatively value or no solution. The following through difference method to calculate under the condition of the Erlang distribution that in order to meet the optimum value should satisfied what kind of conditions. Owing to n^* is the optimum value, there is:

$$(a)L^I(n^*) \geq L^I(n^* - 1)$$
$$(b)L^I(n^*) \geq L^I(n^* + 1)$$

From the first formula we can get that $\frac{n}{f^{-1}(\frac{u}{n})} - \frac{(n-1)}{f^{-1}(\frac{u}{n-1})} \geq \frac{c+k}{s}$, From the second formula we can get that $\frac{n+1}{f^{-1}(\frac{u}{n+1})} - \frac{n}{f^{-1}(\frac{u}{n})} \leq \frac{c+k}{s}$, Combined the two formulas can be as follows:

$$\frac{c+k}{us} \geq \frac{1}{\frac{u}{n^*}f^{-1}(\frac{u}{n^*})} - \frac{1}{\frac{u}{n^*-1}f^{-1}(\frac{u}{n^*-1})} \tag{9}$$

n^* is the point of the first time to meet the value of the above inequality. From the upper formula can be seen the $c + k$ more higher, n^* must also take the greater. Because if you want to make up for the cost. Erlang distribution's order must be increased. The greater the cost and the time of the unit, the Erlang distribution's order smaller can obtain good returns. Also related to the curve change of function $\frac{1}{xf^{-1}(x)}$. In the case of the $L^I - L^P \geq 0$ has solutions is not equal to the optimal solution on the choice of the Erlang distribution system. Because there may be the following three situations.

As can be seen from the picture, when $n^* \geq n_0$. The choice of parallel system is the optimal solution. When $n^* < n_0$, we need to compare the optimal values of the two schemes and the optimal value of the parallel system must also be calculated. It can be solved by the following of differential method. And Its optimal value must meet the following conditions:

$$(a)L^P(n^{**}) \geq L^P(n^{**} - 1)$$
$$(b)L^P(n^{**}) \geq L^P(n^{**} + 1)$$

Similar to the above derivation, the approximate equation is obtained at last:

$$\frac{c}{(1 + \frac{1}{2} + \frac{1}{3} + \cdots + \frac{1}{n})s} \geq \frac{1}{f^{-1}(\frac{u}{n^{**}})} - \frac{1}{f^{-1}(\frac{u}{n^{**}-1})}$$

As can be seen from the formula n^{**} is related to the curve change of function $\frac{1}{f^{-1}(x)}$.

Final comparison, $L^I(n^*)$ and $L^P(n^{**})$, if $L^I(n^*) > L^P(n^{**})$ we need select the Erlang system, On the contrary, choose the parallel system.

4 Conclusion

This paper compared the exponentially distributed and parallel system's reliability with the redundancy backup system. The average lifetime difference of the two systems increases gradually with the increase of the number of parallel systems. The average lifetime difference of the two systems Δs_n, s order is $O(n)$. When you make economic analysis of two systems. If funds are limited, and without considering the economic impact of the failure rate selection on system cost or the cost of the detection switch is too large. Then we should choose the optimal scheme for parallel system. But in consideration of the effect of failure rate selection on the economic cost of the system and when the optimal value point of the Erlang distribution in parallel system and it is at the right side of the intersection point. Then we should select the Erlang distribution system. When the optimal value point is at the left side of the intersection point, Then we should compare the optimal values of the two systems and last choose the best one.

Acknowledgments. This research was partially supported by the School of Mathematics and Computer Science of Fujian Normal University, by the Institute of Innovative Information Industry of Fujian Normal University, by The School of Economic of Fujian Normal University.

References

1. Luss, H.: An inspection policy model for production facilities. J. Manag. Sci. **29**, 101–109 (1983)
2. Bao-he, S.: Reliability and optimal inspection policy of inspected systems. J. OR Trans. **11**(1) (2007)
3. Bao-he, S.: Study on optimal inspection polices based on reliability indices. J. Eng. Math. **25** (6) (2008)
4. Zequeira Romulo, I., Berenguer, C.: On the inspection policy of a two-component parallel system with failure interation. J. Reliab. Eng. Syst. Saf. **88**(1), 99–107 (2005)
5. Xu, L., Xie, C.F., Xu, L.X.: Reliability envelope analysis. J. Comput. **25**(4), 26–34 (2014)
6. Zong-shu, W.: Probability and Mathematical Statistics. China Higher Education Press, Beijing (2008)

Research on the Construction of College Information Applications Based on Cloud Computing

Zhu Quan[1,2(✉)]

[1] Fujian Provincial Key Laboratory of Big Data Mining and Applications,
Fujan University of Technology, Fuzhou 350118, China
loosky@fjut.edu.cn
[2] Center of Modern Education Technology,
Fujan University of Technology, Fuzhou 350118, China

Abstract. The paper focused on the key technology and key issues remain to be solved for the elastic extension mechanism of the IaaS layer and the application layer. Then reformed the system structure of the educational management system, proposed the resource scheduling strategy for the cloud platform, the experimental and operating results showed that the dynamic elastic extension properties of the cloud platform greatly improve the performance of educational management system, can meet the demand in the case of high concurrency. This study can provide a reference for the construction of the college information applications with similar characteristics.

Keywords: Cloud computing · College informationization · Information technology applications · Dynamic elastic expansion · Resource scheduling strategy

1 Introduction

To solve the problem of the massive data processing, Amazon, Google and other companies put forward the concept of cloud computing in 2006 [1, 2]. It has gained extensive concern of the industry and academia, Governments have also regard cloud computing as a national strategy, invested a lot of financial and material resources to the study of cloud computing technology and the deployment of the cloud platform and applications.

The feature of elastic extension, resource pooling and on-demand for the cloud computing, It is very suitable for the application environment which data elastically changed, such as taobao's "double 11" big promotion and 12306 booking website. During the Spring Festival of 2015, the peak of 12306 web site reached 29.7 billion times, the key to solve this problem was build a scalable cloud platform architecture, the resource can be deployed at any time according to the need of the business, then ensure the sustainability of business under high pressure.

Now the colleges and universities have entered in the process of the construction of "intelligent campus", is designed to a variety of application service system as the

J. Pan et al. (eds.), *Intelligent Data Analysis and Applications*, Advances in Intelligent Systems and Computing 535, DOI 10.1007/978-3-319-48499-0_6

carrier, fully mix the teaching, scientific research, management and campus life, thus provide a more humanized campus information services for school teachers, students and alumnus. Therefore, the paper will research the key technologies for cloud computing such as elastic extension, and take educational administration management system as an example, discussed the construction model of college campus informatization under the condition of the cloud computing.

2   Cloud Computing and University Informationization

Cloud computing reflect the application and service innovation for the information technology, it is also the important direction for the development of information technology industry, and could promote the innovation of the economic and social development. The state council issued the "State council on promoting the cloud innovation and development to foster new forms of information industry" in January 2015, the ministry of education issued the "development plan of education informatization decade (2011–2020)", it puts forward the specific development goals, by fully integrate existing resources, using cloud computing technology, forming intensive development way of the allocation of resources and services, build a new model of college informationization development.

Along with the increase in the number of the university informationization applications as well as the continuous increase of the utilization level of informatization, the data resources also is in rapid growth, but because of the general lack of unified management norms, the difference of data format, the business department to build their own application systems, and many other problems, make large amounts of data can not be effectively sharing and utilization of various information between applications and there are many "information islands". Through the cloud computing technology to build the unified data processing platform, to realize data information sharing and collaboration, and through the large data mining technology to realize dynamic and real-time processing of mass information data, thus provide all kinds of information services for the management of colleges and universities and the teachers, students. At the same time by using technologies such as virtualization, cloud computing platform integrated servers, storage, network and other hardware resources, optimize allocation of system resources, providing the elastic extensible platform for the big data processing, thus realized the flexibility of application deployment. It could greatly improve the resource utilization, reduce the total energy consumption and operational costs [3, 4].

Therefore, how to integrate the university informationization applications and the processing, analysis, mining and utilization of the education informationization big data, will be the key of the future development for the university informationization, and cloud computing technology with advantages of the automation scheduling for IT resource and rapid deployment, and the elastic extension, will become an important technical means to solve this problem [5, 6].

3 The Key Technology of Dynamic Elastic Extension for Cloud Platform

The cloud computing technology will bring a lot of advantages, first of all, the ability of data storage and maintenance provides storage and processing power for campus big data. Secondly, by using of distributed computing methods which can achieve high performance parallel processing ability, It provides safeguard for the campus informationization applications under high concurrency and the analysis services of campus based on data mining. Thirdly, cloud computing provides elastic extension ability which can ensure the high availability of campus informatization and the expansibility, thus meet the user's access and the experience.

For the critical elastic extension technology, mainly including vertical scaling (upgrade the nodes such as CPU, memory) and horizontal scaling (set up or destroy nodes). For how to make use of the advantage of cloud computing technology to the construction of campus informatization, then realize the elastic extension of campus informatization, may require the comprehensive consideration from the IaaS layer, middleware layer and application layer of the campus informationization cloud [7].

4 Construction of Educational Administration System Based on Cloud Computing

In the campus informationization applications, the role of educational administration system is important, it is responsible for the school curriculum, student registration, teachers manage the examination result, student achievement query and a series of functions, is the foundation of the school teaching work and the construction of digital campus, its operation is stable or not directly affects the teaching activities of the entire teachers and students. And the operation of the educational administration management system, also have certain characteristics: most of the ordinary time, the teachers and students is mainly for conventional query of score and schedule, the number of concurrent connections does not too high, only need a small number of server resources to support the access requirements. But during the students select the course, there will be a high concurrency, often can make the abnormal phenomenon such the system down or denial of service.

The paper would upgrade the educational administration management system based on cloud computing technology, realized the normal access under high concurrent environment of educational administration system through the dynamic elastic extension of the cloud computing, and in the usual off-peak, cloud resources can be recycled and used for the other campus informatization applications, thus improve the resource utilization efficiency of the whole campus informatization cloud platform.

4.1 The System Architecture of Educational Administration Management

The logical architecture of educational administration system which the author's school used shown as Fig. 1, the software adopted the B/S + C/S architecture, which the Web and APP layer deployed in Windows IIS service. It actually with two layer architecture and logically for three layers.

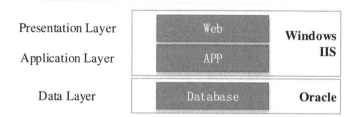

Fig. 1. The logic structure diagram of educational administration management system

The system function including teacher management, student services and database. The actual deployment architecture of the system is shown in Fig. 2.

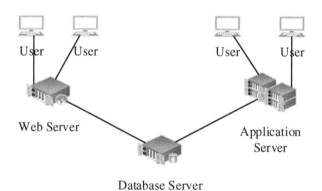

Fig. 2. The architecture diagram of educational administration management system

Through the analysis for the system software architecture of the educational administration management system, its performance bottleneck mainly produces in the front-end Web presentation layer and database layer. And through the actual test found that the access bottleneck mainly lies in the Web server on the number of simultaneous connections limit. The front-end Web presentation layer provides the user interface, when lots of students select the courses which cause a number of simultaneous connections, the system will become deadlock or inaccessible due to the system performance limit. Therefore, the Web layer can use cloud computing technology, through the dynamic elastic extension ability, when facing the large traffic, the system will

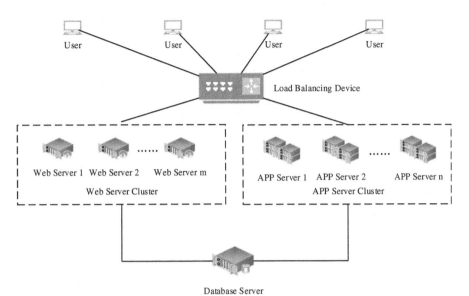

Fig. 3. The architecture diagram of educational administration management system based on Cloud Computing

improve the concurrent traffic by adding server nodes. At the same time, through the load balancing device on each node server to balance the load. Based on this, the improved system architecture of educational administration management system based on cloud computing as shown in Fig. 3.

4.2   The Experimental Test

The paper takes a series of experiments based on the system architecture is shown in Fig. 3, the related product selection of the platform is in the following Table 1:

Table 1. The system function selection for the cloud platform

Function	Product
Computing resource	Blade server
Resource pooling	H3C CAS
DataBase middleware	Oracle 10g
Load balancing device	H3C L5000-C

The components of H3C CAS can realize the monitoring for the operating system, middleware and applications. The paper set up the monitoring items include: CPU usage (CPU, %), memory usage(memory, %) and the number of simultaneous connections (link_connections), monitoring frequency set to 30 s. Due to the number of simultaneous connections are the main factors influencing the user access to the

Table 2. The resource scheduling strategy for cloud platform

The resource scheduling strategy for cloud platform

Step 1 :   Initialization, respectively set the monitoring threshold to start node as **cpu_A**, **memory_A**, **link_connections_A**, *and the threshold to close nodes as* **cpu_B**, **memory_B**, **link_connections_B**.
    The resource pool start one node by default, and set the node is not added to the resource scheduling strategy, which prevent nodes were shut down.
Step2 :   if cpu > cpu_A or memory > memory_A or link_connections > link_connections_A then
            Set up a new virtual machine node;
            Start the virtual machine node;
Step3 : else if cpu < cpu_B or memory < memory_B or link_connections < link_connections_B then
            Shut down the virtual machine node;
            Delete the virtual machine node;

educational administration management system experience, thus can be set up on load balancing device based on network access number of connections to a cloud platform for load balancing.

The paper design the resource scheduling strategy for cloud platform is shown in the following Table 2.

In the actual situation, due to the long time to set up a new virtual machine in the step2, It may copy a certain number of virtual nodes in advance, and the delete operation for virtual machine nodes in step3 may not execute.

The paper takes a series of experimental tests for the cloud platform and the resource scheduling strategy of cloud platform by the pressure test software-LoadRunner, and the running environment of administration management system is deployed on the cloud platform. The paper set a single Web server node as 4 vCPU, 4 gb memory, and simulate the students select the courses by LoadRunner, test the scene of 16 students login per second at the same time and continuous 6000 students to log in. Test results as shown in Fig. 4.

It can be seen from the diagram, between the 30 s and 40 s, there is a wave crest of the response time, which mean the login response time will increases with the increasing number of the students. Due to the number of connections for resources expansion achieve the threshold, the cloud resource scheduling strategy launched a second virtual machine node. After the second node starts, the response time of the system fall fast and have a steady state. In actual production environment, the cloud platform support the 7000 online students to select courses at the same time by 20 Web server.

The improved educational administration system based on cloud computing technology, It can extend the virtual machine nodes real-time and dynamically through the cloud platform resource scheduling strategy, then ensure the seamless performance

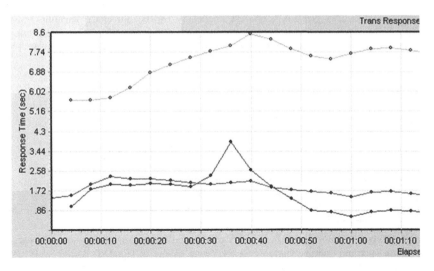

Fig. 4. The pressure test results for Cloud platform

expansion for the educational administration system, and provides a good experience for the entire teachers and students. Also the cloud resources can realize automatic recovery, in spare time, the educational administration system resources can be released automatically and provide for other applications, thus greatly improving the efficiency of the cloud platform.

5 Conclusion

Based on the specific needs of cloud computing technology for the informatization construction of colleges and universities, the paper discussed the key technologies to realize dynamic elastic extension for cloud platform respectively from the IaaS layer, middleware layer and application layer. The paper reform the system architecture of educational administration management system based on cloud computing technology, put forward the cloud resource scheduling strategy, and take the related test, the test results and actual running results show that the dynamic elastic extension character-istics of cloud computing greatly improve the performance of educational adminis-tration management system, meet the application requirement of the system under the condition of high concurrency.

Except the educational administration management system, there is a series of campus informatization applications with similar features, such as web site group system, enrollment system in the usual load will not too high, and in some special time such as admissions, celebration, its visitors will surge. Therefore, the research in this paper, also can provide a reference for the construction of related campus informati-zation applications.

Acknowledgment. This work is supported financially by the university scientific research special of Fujian Province (No. JK2014033), Special of college education informationization in education department of FuJian province (No. JA15326), Dr. Startup funds of FJUT (No. GY-Z15009).

References

1. Jun-Zhou, L., Jia-Hui, J., Ai-Bo, S., et al.: Cloud computing: architecture and key technologies. J. Commun. **32**(7), 3–21 (2011)
2. Fumin, Z., Jiang Xinhua, H., Huichun, H., et al.: A review on the research and application of cloud computing. J. Fujian Univ. Technol. **11**(03), 231–242 (2013)
3. Yi-Jie, W., Wei-Dong, S., Song, Z., et al.: Key technologies of distributed storage for cloud computing. J. Softw. **23**(4), 962–986 (2012)
4. Bo, Y., Ying, W., Luo-Ming, M., et al.: A new virtual machine migration strategy based on migration cost and communication cost for power saving in cloud. J. Beijing Univ. Posts Telecommun. **35**(1), 68–71 (2012)
5. Xiong-Pai, Q., Hui-Ju, W., Fu-Rong, L., et al.: New landscape of data management technologies. J. Softw. **24**(02), 175–197 (2013)
6. Shen, D.R., Yu, G., Wang, X.-T., et al.: Survey on NoSQL for management of big data. J. Softw. **24**(08), 1786–1803 (2013)
7. Quan, Z.: Research on the key technology of dynamic elastic expansion for cloud platform **14** (01), pp. 71–75 (2016)

Estimating Human Activities from Smartwatches with Feedforward Neural Networks

Sebastián Basterrech[(✉)]

Faculty of Electrical Engineering and Computer Science,
Department of Computer Science, VŠB-Technical University of Ostrava,
Ostrava, Czech Republic
`Sebastian.Basterrech.Tiscordio@vsb.cz`

Abstract. A key task in the Human Activity Recognition area consists in defining a Machine Learning Classifier. We analyse an approach for modelling the activities using Neural Networks. We consider feedforward networks applied for temporal learning, therefore the network inputs collect information from the past. The input patterns cover information on a time-range window of the past activities, as well as external variables. We evaluate our approach on a well-known dataset and we compare our results with the obtained results in the literature.

Keywords: Human Activity recognition · Neural Network · Classification · Temporal learning

1 Introduction

Human Activity Recognition (HAR) is a multidisciplinary research area composed by the expertise and knowledge from several including: computer-science, cognitive sciences, sociology and psychology. HAR is devoted to understand the behaviour and collective dynamics of humans. Nowadays, we dispose of a huge number of devices for recording information about humans and their environments. During the last decades, there have been important developments of sensors technologies, video cameras, audio recording devices and other tools. They have had a significant growth in diversity and quality. In addition the costs of those devices have rapidly declined over the years, and now is possible to collect valuable information in real time and at low cost. As a consequence, at the present we have many opportunities for studying problems concerning to the human behaviour, and for developing the HAR area.

In many countries the use of mobile phones covers a huge percentage of their population, and the number is increasing in societies with low development. Even though the smart watches are less popular, they also are used for a huge amount of persons around the globe and its trend is also increasing over the years. Hence, several articles have been presented in the community, in which a machine learning model is used for mapping the collected information by the sensors, smart phones, etc., and the human activities [1]. Most often, the researchers

© Springer International Publishing AG 2017
J. Pan et al. (eds.), *Intelligent Data Analysis and Applications*, Advances in Intelligent Systems and Computing 535, DOI 10.1007/978-3-319-48499-0_7

have used a classical statistic learning classifier for recognising classes, where each class represents a human activity. If the information is collected in a table where its columns represent the features and its rows represent events on the time, the most applied approaches have the hypothesis that the rows are independent each of the other one. For instance, in [2] the authors applied the following learning techniques K-NN, SVM, Random Forest, C4.5. Other recent works where SVM were used [3–5], as well as Neural Networks (NNs) works were applied in [6,7].

This is a short article where is presented our preliminary work on the HAR area. In particular, we use a NN as a machine learning classifier for the human activities. A difference of our approach with previous works, it is that the input pattern contains two type of information: the current multidimensional event and the past. Our work hypothesis is that the current human activity is dependent of both the current and precedent human activities. Therefore, we are modelling the problem using temporal learning methods. The novelty of our approach is the consideration of time as important variable for making the classifier. We can see human activities as actions evolving in time, then the NN should learn the spatial information contained into the input variables (given by the sensors) as well as the temporal dependencies among the events (given by precedent activities). In [2] the authors created a rich dataset containing activity information of several humans from several devices. Another advantage of this dataset is that is public, therefore we can compare our results with the available results on the community.

The rest of the article is structured as follows. In the next section we present a brief background of the classification problem using temporal information and we revisit the feedforward neural networks. Then, we present some literature in the area of HAR. Section 3 contains the description of the dataset and our empirical results. As usual, we finish with a global discussion about our contribution and future work.

2   Methodology

Given a dataset composed by input-output pairs where the outputs are labels, the goal in classification consists in to make a mapping between the inputs and the labels (classes). The mapping model should be able to assign labels to input patterns "*as well as possible*" using the information of the given dataset. In this study, we don't assume independence among the input patterns. More formal, given a learning dataset S composed of the pairs $(\mathbf{x}(t), y(t))$ with $\mathbf{x} \in A^n$ and $y \in \mathbb{R}$ with $y(t) \in \{C_1, C_2, \ldots, C_M\}$, the goal consists in defining a parametric learning model $f(\theta, \cdot)$ such that it is able to label the sequences in a test set S' with $S \cap S' = \emptyset$ as accurate as possible. We are using the quadratic distance among the targets and the predicted values for estimating the accuracy of the learnt model. We denote the predicted class at time t by $\hat{y}$, so the distance is:

$$E = \frac{1}{T} \sum_{t=1}^{T} (\hat{y}(t) - y(t))^2, \tag{1}$$

where T is the number of seen training pairs and $\hat{y}(t)$ is the predicted output of $\mathbf{x}(t)$. We are assuming that the outputs are unidimensional that is the case of our study, although the generalisation to other dimensions is straightforward.

2.1 Feedforward Neural Network as Human Activity Recognition Tool

A NN is a parallel distributed system composed by a set of simple processors called neurons interconnected among them. In the *Feedforward Neural Network (FNN)* case, the neurons are connected following an acyclic graph, they are arranged into interconnected layers. There are many applications of FNNs for solving supervised learning problems, due to its good performance in some real-world applications, and that the training process is faster that network with recurrences. The parameters of FNNs with "few" layers are easier to set up than the parameters of recurrent networks. The model parameters are the weight connections between the neurons, they are adjusted in order of generating the mapping input-output according the given dataset. One the most used algorithm for adjusted the weights is the back-propagation method, which is optimization gradient based type algorithm [8].

More formal, we can see a FNN as a parametric function from an input space A^n to an output space B^m, its multidimensional parameter is composed by matrices W_1, W_2, W_{l-1} that collect the weights and vectors $\mathbf{b}_1, \mathbf{b}_2, \mathbf{b}_{l-1}$ named *bias*. In the following we simplify and we present the model such that A and B are real spaces, and B is unidimensional. The network function is computed by the following recurrent sequence:

$$\mathbf{z}_{i+1} = s_i(W_i \mathbf{z}_i + \mathbf{b}_i), \tag{2}$$

where initial expression is given by $\mathbf{z}_1 = \mathbf{x}$ and $\mathbf{z}_l$ contains the output of the FNN, then $\hat{y} = \mathbf{z}_l$. The function $s_i(\cdot)$ is often selected sigmoid or $\tanh(\cdot)$. The weights are computed using a learning dataset and optimisation algorithms in order of minimising the expression (1). The standard approach for applying FNNs for solving temporal learning problems consists in using a network with l input neurons, arbitrary number of hidden neurons, and one output neuron (if the target is unidimensional). The number of input neurons l is composed by the information collected at current time, and arbitrary selected information collected at precedent times. If at each time t, the input pattern has dimension n, and we arbitrary selected q previous activities, then the number of inputs of the network will be $l = n + q$. Note that, in real problems when we are predicting activities we know the real value of the estimated activity after some delay Δ_t. This delay can be few seconds or minutes, as well as it can be many days, its value depends of the specific problem. In the case that Δ_t is large, the values of the q precedent activities should be given by the estimated values of the network. In the case that Δ_t is small, then we can use the exact precedent activities in a relative short time range. Therefore, it depends of the problem how to make the network and how to evaluate it. In this preliminary work, we are using Δ_t equal to 50 times steps, this means that the network can use information of previous activities only if they were done at least 50 time steps before the current time t. In the time range $[t - \varDelta_t + 1, t]$ we can use only estimations of the human activities.

3   Experimental Results

3.1   Description of the Dataset

We apply FNNs on a large dataset presented in [2]. The main goals studied in [2] were to analyse the impact caused for collecting data from several heterogeneous sensors over the classifiers performance. On the other hand, in this first paper we are interested only in making a good classifier and if the classifier construction using time overcomes the results with respect of those obtained when the time is not considered. The data was collected with an embedded Gyroscope sensor, which was sampled at the highest frequency, the tools used for collected the data were two LG smartwatches and two Samsung Galaxy Gears smartwatches [2]. The activities of three subjects were measured, and the considered activities were: Biking, Sitting, Standing, Walking, Stair Up and Stair down. We encode the activities in $[0, 1]$ as: $\{0, 1/6, 1/3, \ldots, 1\}$. The input patterns are the coordinates x, y, z, the subject, the watch model, and the device. More details about the benchmark dataset are available in [2]. Figure 1 shows three graphics with

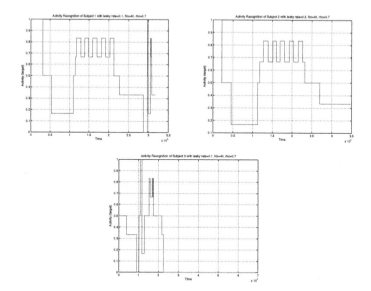

Fig. 1. The three figures show the activities of the subjects on the testing dataset. At the left top corresponds to the subject 1, right top corresponds to the subject 2 and on the bottom corresponds to the subject 3. The horizontal axis corresponds to the time, the vertical one represents the activity.

the activities according to the time of each subject. We can see very different behaviour between the subjects.

3.2   Results

In this short article, with our work still in progress, we evaluate the approach using two types of feedforward network topologies. One topology consists in two hidden layers of 20 neurons each one, another one consists of 3 layers each one with 30 neurons. The applied optimisation algorithm for finding the weights is Levemberg-Marquardt, and we are using the NNtoolbox of Matlab. We have three subjects, we have a training and a validation dataset for each subject. Due to the large amount of points, we use a third part of the dataset for training and the rest for evaluating the model. For subject A we use 136083 points for training and 317527 points for testing, for subject B we use 149204 points for training and 348144 points for testing, and for the subject C we use 29285 samples for training and 68332 for testing. Table 1 shows the MSE obtained for the FNN with different topologies. The first two rows corresponds to a non-temporal modelling, it means the inputs consist only of the measurement of current time. The last two rows shows the MSE using temporal modelling, we compose the input pattern with the coordinates x, y, z, the device and previous activities of the subject, we arbitrary selected the activities at the following previous times: $\{t - 50, t - 100, t - 200, t - 300, t - 500\}$. Figure 2 present the targets and the predictions on the training dataset when a FNN with a $9 : 30 : 30 : 30 : 1$ topology. We can see how the predictions approach very well the target. Even the estimation can be better, if we consider that the outputs are discreet, then we can round the real outputs to the discreet values. On the other hand, Fig. 3 shows

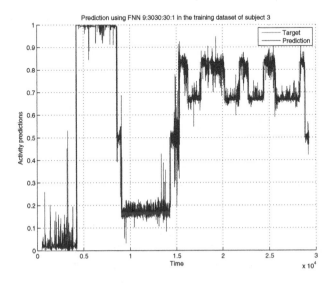

Fig. 2. Prediction versus targets on the training dataset of subject C.

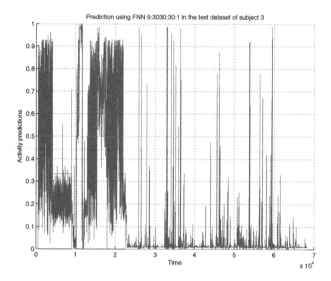

Fig. 3. Prediction versus targets on the test dataset of subject C.

the predictions on the testing dataset, we can see that the error is increased, also is possible to see how the learning machine can estimate better some activities than other ones.

Table 1. MSE using different topologies of the FNN

	Topology	Subject A	Subject B	Subject C
Non-temporal modeling	4:20:20:1	0.2268	0.2629	0.1936
	4:30:30:30:1	0.1548	0.2266	0.2124
Temporal modeling	9:20:20:1	0.0203	0.0460	0.0696
	9:30:30:30:1	0.0210	0.0364	0.0124

4   Conclusions and Future Work

Here we present a preliminary study about the application of FNNs for recognising human activities. Basically, the main idea of the paper consists in solving the classification problem considering historical information of the activities, instead of using only the information at the current time. In the near future, the research lines include the following topics. To perform experiments with other types of Neural Networks. We would like to modify the topology as well as the activation functions. In this paper the evaluation of the classifier is using MSE, it will be interesting to use another type of accuracy such as Cross-Entropy and F-Score. How to define the input pattern containing information of the past is a hard task, in this way we would like to test several time-windows with past information as input pattern in order to find better solutions. Besides, a similar approach can be applied on other family of classifiers. We would like to compare the performance of FNNs with other popular classifiers.

Acknowledgement. This work is supported by Grant of SGS No. SP2016/68 and SP2016/97, VŠB–Technical University of Ostrava, Czech Republic.

References

1. Lara, O.D., Labrador, M.A.: A survey on human activity recognition using wearable sensors. IEEE Commun. Surv. Tutorials **15**(3), 1192–1209 (2013)
2. Stisen, A., Blunck, H., Bhattacharya, S., Prentow, T.S., Kjaergaard, M.B., Dey, A., Sonne, T., Jensen, M.M.: Smart devices are different: assessing and mitigating mobile sensing heterogeneities for activity recognition. In: Proceedings of 13th ACM Conference on Embedded Networked Sensor Systems (SenSys 2015), Seoul, Korea (2015)
3. Mannini, A., Sabatini, A.M.: Machine learning methods for classifying human physical activity from on-body accelerometers. Sensors **10**(2), 1154–1175 (2010)
4. Anguita, D., Ghio, A., Oneto, L., Parra, X., Reyes-Ortiz, J.L.: Human activity recognition on smartphones using a multiclass hardware-friendly support vector machine. In: Bravo, J., Hervás, R., Rodríguez, M. (eds.) IWAAL 2012. LNCS, vol. 7657, pp. 216–223. Springer, Heidelberg (2012). doi:10.1007/978-3-642-35395-6_30
5. Kwapisz, J.R., Weiss, G.M., Moore, S.A.: Activity recognition using cell phone accelerometers. ACM SigKDD Explor. Newslett. **12**(2), 74–82 (2011)
6. Yang, J.-Y., Wang, J.-S., Chen, Y.-P.: Using acceleration measurements for activity recognition: an effective learning algorithm for constructing neural classifiers. Pattern Recogn. Lett. **29**(16), 2213–2220 (2008)
7. Jeroudi, Y.A., Ali, M.A., Latief, M., Akmeliawati, R.: Online sequential extreme learning machine algorithm based human activity recognition using inertial data. In: 2015 10th Asian Control Conference (ASCC), pp. 1–6. IEEE (2015)
8. Bishop, C.M.: Neural Networks for Pattern Recognition. Oxford, New York (1995)

Image Processing and Applications

Image Classification Based on Image Hash Convolution Neural Network

Yaoxing Chen, Yunyi Yan[⊠], and Dan Zhao

School of Aerospace Science and Technology,
Xidian University, Xi'an 710071, China
alice_chena@126.com, yyyan@xidian.edu.cn

Abstract. In image classification tasks, in order to improve the classification accuracy, we need to extract the multidimensional sample characteristics. However, for massive amounts of data, calculation and storage is a big bottleneck. In this paper, a method named image hash was proposed to solve this problem. Image hash can code high-dimensional image feature for simple binary code. Feature extraction is the most important step in image hash. In the current works about image hash, feature extraction needs artificial experience to design feature extractor, which is complicated and not intelligent. Convolution neural network can take original image as input to obtain from the bottom level to the top level of characteristics, which is robust for translation zooming and rotation etc. Therefore, this paper proposes the image hash combined with convolution neural network for image classification, and the experiment proves that it has good classification effect.

Keywords: Convolutional neural network · Image hash · Dropout · Deep learning

1 Introduction

In the field of object classification, the most important thing are feature extraction and representation abstraction. Traditional object recognition methods need to design the feature extraction based on experience [9], deep learning can bypass this difficulties because the network learn the inherent characteristics of the images. LeNet-5 [4], one of the most classical convolutional neural network [2], consist of multiple layers to learn multi-layer abstract representations of the data. As mentioned in the literature [3], the most obvious feature of the image hash algorithm [5] is robustness, namely an image is the same or similar with its abstract (also called a hash list) which is gotten by image hash algorithm. A typical image hash algorithm is constituted of the preprocessing, quantification and match.

We put forward a model of image hash algorithm based on LeNet-5. We get the fingerprint of images using image quantization operation of hash algorithm on the characteristics extracted by convolutional neural network, and we train the whole

This work was supported by the National Natural Science Foundation of China under Grants No. 61571346 and No. 61305041.

J. Pan et al. (eds.), *Intelligent Data Analysis and Applications*, Advances in Intelligent Systems and Computing 535, DOI 10.1007/978-3-319-48499-0_8

network including feature extraction and hash mapping function by supervised learning to get a better image classification.

2 Dataset

ORL (Olivetti Research Laboratory) face data is a face dataset in which the face images were taken between 1992 and 1994 made by Olivetti research institute of the University of Cambridge. The data set includes 400 images, 40 individuals, each 10 images. Each person in the face image taken in different time, so they are different in light and some facial details. All images are based on a black background, and the image face towards the front. The original image resolution is 112×92. In order to reduce the network parameters, and the training time, we resize it to 57×47.

3 Architecture

Our network is modified based on the traditional LeNet-5. As shown in Fig. 1, our network begins with a 57×47 input image, which are used to encode the pixel intensities for the face image. This is then followed by first convolutional layer with 6 kernels of size 5×5 with a stride length of 1 pixel. The result is a layer of $6 \times 53 \times 43$ hidden feature neurons. The next step is a max-pooling layer, applied to 2×2 regions, across each of the 6 feature maps. The result is a layer of $6 \times 26 \times 21$ hidden feature neurons. We consider convolutional and max-pooling layer as a whole convolutional layer. The second convolutional layer takes as input the out of the first convolutional layer(including pooling layer)and filters it with 20kernels of size $5 \times 5 \times 6$. The final layer of connections in the network is a fully-connected layer. The third convolutional layer has 50 kernels of size $5 \times 5 \times 20$ connected to the outputs of the second convolutional layer(including pooling layer). The fully-connected layers have 1000 neurons and take as input the flatten form of the previous layer's output. Our output of full connection layer is quantified mapping to the hash code which is taken as input of classifier. To reduce overfitting, in the fully-connected layers we employed a recently-developed regularization method called "dropout" that proved to be very effective.

Convolutional layers use three basic ideas: local receptive fields, shared weights, and pooling.

3.1 Local Receptive Fields

In a convolutional net, we think the input as a *height* × *width* square of neurons, whose values correspond to the *height* × *width* pixel intensities in input image.

And then, we'll connect the input neurons to hidden neurons of the next layer. But we won't connect every input pixel to every hidden neuron. Instead, we only make connections in small, localized regions of the input image. That region in the input image is called the local receptive field [4] for the hidden neuron. Each connection learns a weight. And the hidden neuron learns an overall bias as well.

We then slide the local receptive field across the entire input image by one pixel at a time. For each local receptive field, there is a different hidden neuron in the hidden layer. In fact, sometimes a larger stride length [6] is used for some high resolution image.

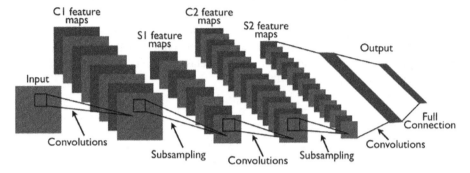

Fig. 1. It consists of four learned layers- three convolutional and one fully-connected followed by a classifier of Logistic Regression.

3.2 Shared Weights and Biases

Each hidden neuron has a bias and 5×5 weights connected to its local receptive field. And we use the same weights and bias for each of the hidden neurons. In other words, for the j, k th hidden neuron, the output is:

$$\sigma\left(b + \sum_{l=0}^{4} \sum_{m=0}^{4} W_{l,m} a_{j+l,k+m}\right) \tag{1}$$

Here, σ is the neural activation function - perhaps the sigmoid function, or tanh, or ReLUs, b is the shared value for the bias. $W_{l,m}$ is a 5×5 array of shared weights. And, finally, we use $a_{x,y}$ to denote the input activation at position x, y. This means that all the neurons in the first hidden layer detect exactly the same feature, just at different locations in the input image. We call the weights defining the feature map the shared weights or convolutional kernels. And we call the bias defining the feature map the shared bias. A great advantage of sharing weights and biases is that it greatly reduces the number of parameters involved in a convolutional network.

To do image classification we'll need more than one feature map. And a complete convolutional layer consists of several different feature maps.

3.3 Pooling Layers

Convolutional layers also contain pooling layers [2]. What the pooling layers do is simplify the information in the output from the convolutional layer. As a concrete example, one common procedure for pooling is known as max-pooling. In max-pooling, a pooling unit simply outputs the maximum activation in the 2×2 input region. As mentioned above, the convolutional layer usually involves more than one single feature map. We apply max-pooling to each feature map separately.

3.4   Logistic Regression

Logistic regression is a probabilistic, linear classifier. It is parametrized by a weight matrix W and a bias vector b. Classification is done by projecting data points onto a set of hyper planes, the distance to which reflects a class membership probability.

Mathematically, this can be written as:

$$P(Y = i|x, W, b) = soft\,\max_i(W_x + b)$$
$$= \frac{e^{W_i x + b_i}}{\sum_j e^{W_j + b_j}} \qquad (2)$$

The output of the model or prediction is then done by taking the maximum of the vector whose i'th element is $P(Y = i|x)$.

$$y_{pred} = \arg\,\max_i P(Y = i|x, W, b) \qquad (3)$$

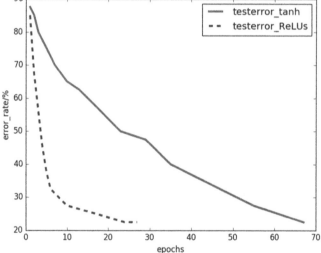

Fig. 2.   The test errors of tanh activation function and ReLUs activation function

3.5   Using Rectified Linear Units

The LeNet-5 uses a sigmoid activation function [4]. Following Nair and Hinton [11], we refer to neurons with this nonlinearity as Rectified Linear Units (ReLUs). Deep convolutional neural networks with ReLUs train several times faster than their equivalents with tanh units, which is shown is Fig. 2. ReLUs have a property that they do not require input normalization to prevent them from saturating.

3.6   Expanding the Training Data

The original dataset is divided in 200 examples for the training set, 120 examples for the validation set and 80 examples for testing. Our data set is so small that our network might be overfitting, and can't get good accuracy on the testing set. We algorithmically expand the training set to 1000. A simple way of expanding the training data is to displace each training image by a single pixel, either up one pixel, down one pixel, left one pixel, or right one pixel, as is shown in Fig. 3. In fact, expanding the data turned out to considerably reduce the effect of overfitting.

Fig. 3. The first line shows ten pictures from the expanded training set. The second line shows ten pictures from the validation set. The third line shows ten pictures from testing set.

3.7   Dropout

We use stronger regularization techniques—dropout [7] to reduce the tendency to overfitting. The basic idea of dropout is to remove individual activations at random while training the network. This makes the model more robust to the loss of individual pieces of evidence, and thus less likely to rely on particular idiosyncracies of the training data and so the net learned faster.

We only applied dropout to the fully-connected layers, not to the convolutional layers. There's no need to the convolutional layers, because the shared weights helps convolutional filters to learn from across the entire image. This makes them less likely to pick up on local idiosyncracies in the training data.

3.8   Image Hash

Image hash method is mapping high-dimensional data in Euclidean space to lower dimensions of binary space. Assume that the input is a n-dimensional vector $X = [x_1, x_2, \cdots, x_n]$, where, x_i is a m-dimensional sample. If the expected binary space dimension is b, we need to define a hash function $\{h_j, \cdots h_b\}$. Each hash function h_j is a binary function, b hash function can map a m-dimensional object x_i into b-dimensional binary code $H_i = \{0,1\}^b$, which is called hash code. Because it can make compact binary code for high-dimensional image feature vector, image hash method has good performance of the characteristics of storage and speed.

In our model, we use image hash to map the output of fully-connected layer to binary code. The output of fully-connected layer represents the feature of the original input. Though the hash code, we can transform the feature into a binary code. In the following layer, we take the hash code as the input of classifier.

4   Experiments

4.1   Details of the Network Learning

We build our deep convolutional neural network and ran it using theano. Theano is a python library that makes writing deep learning models easy, and gives the option of training them on a GPU. We trained our model using Minibatch Stochastic Gradient Descent with a batch size of 40, learning rate of 0.001 (Figs. 4, 5 and 6).

4.2   Features Extracted from the Network

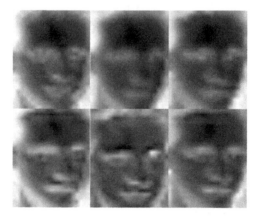

Fig. 4. The six feature maps extracted by the first convolution layer.

4.3   Results

We test our deep convolutional neural network on the dataset we introduced in Sect. 2. We achieved test error rate of 2.5 %. For comparison, we also are tested on other typical model respectively. The results are shown in Table 1.

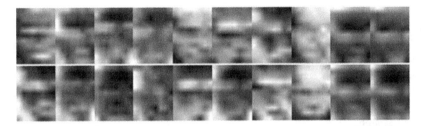

Fig. 5. The twenty feature maps extracted by the second convolution layer.

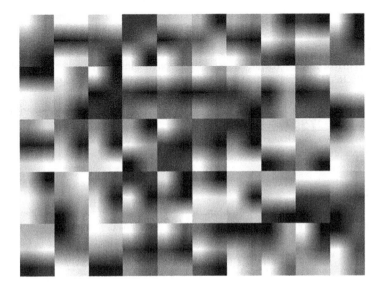

Fig. 6. The fifty feature maps extracted by the third convolution layer.

Table 1. Comparison of the face classification errors on the ORL dataset.

Network	Testerrorrate
CNN_1	5 %
SDA	37.5 %
Shallow	40 %
CNN*	*2.5 %*

CNN_1 has three convolutional layers and fully-connected layer the same as our model, but without hash coder. SDA is stack denoising autoencoder with three hidden layer pre-trained and fine-tuning afterword. Shallow is a shallow network with fully connected layers. *CNN** is our deep convolutional neural network combined with image hash.

5 Conclusion

In this paper, we proposed a fusion model of convolutional neural network and image hash and got good classification effect in our dataset. Because there is only 80 images in our test set, we can achieve an error rate of 2.5 %, namely only two images were misclassified. We can use bigger dataset in later work. Image hash was widely used in image retrieval. In the future work, we can apply our model to image retrieval for the further research.

References

1. LeCun, Y., Bengio, Y., Hinton, J.: Deep learning. Nature **521**, 436–444 (2015). 05/28/print
2. Dahl, J.V., Koch, K.C., Kleinhans, E., et al.: Convolutional networks and applications in vision. C. In: IEEE International Symposium on Circuits & Systems, pp. 253–256. IEEE Press (2010)
3. Zhaoqing, L., Qiong, L., Jingrui, L., Xiyuan, P.: SIFT based image hashing algorithm. Chin. J. Sci. Instrum. **32**(9), 2024–2028 (2011)
4. LeCun, Y., Boser, B., Denker, J.S., Henderson, D., Howard, R.E., Hubbard, W., Jackel, L.D.: Backpropagation applied to handwritten zip code recognition. Neural Comput. **1**, 541–551 (1989)
5. Fu, H.: Research on Hashing—Based Image Retrieval for Large—ScaleDataset. Dalian University of Technology (2014)
6. Ciresan, D.C., Meier, U., Masci, J., Maria Gambardella, L., Schmidhuber, J.: Flexible, high performance convolutional neural networks for image classification. In: IJCAI Proceedings-International Joint Conference on Artificial Intelligence, p. 1237 (2011)
7. Krizhevsky, A., Sutskever, I., Hinton, G.E.: Imagenet classification with deep convolutional neural networks. In: Advances in Neural Information Processing Systems, pp. 1097–1105 (2012)
8. Ciresan, D.C., Meier, U., Schmidhuber, J.: Multi-column deep neural networks for image classification. In: 2012 IEEE Conference on Computer Vision and Pattern Recognition (CVPR), pp. 3642–3649 (2012)
9. Fu, Y., Wang, Y.W., Wang, W.Q., Gao, W.: Content-based natural image classification and retrieval using SVM. Chin. J. Comput. **26**, 1261–1265 (2003)
10. Lecun, Y., Bottou, L., Bengio, Y., et al.: Gradient-based learning applied to document recognition. J. Proc. IEEE **86**(11), 2278–2324 (1998)
11. Nair, V., Hinton, G.E.: Rectified linear units improve restricted boltzmann machines. In: Proceedings of 27th International Conference on Machine Learning (2010)

Object Recognition Based on Superposition Proportion in Binary Images

Yan Zheng, Baolong Guo, and Jing Ma$^{(\boxtimes)}$

School of Aerospace Science and Technology, Xidian University,
Xi'an 710071, China
blguo@xidian.edu.cn, mjy_9527@126.com

Abstract. This paper describes an approach for recognizing object. Object recognition plays an important role in intelligent system. But usually the existing methods can hardly deal with occluded, shadowy and blurred images. A novel approach is proposed in this paper. This proposed approach based on prior model recognize the main regions and direction of object to adjust the model on size and direction. Finally the proposed approach compute the superposition proportion similarity, between model and object. The experimental result shows that this proposed approach is robust to scenarios containing occlusion, cluttering and low contrast edges.

Keywords: Object recognition · Occluded object · Prior model · Binary images

1   Introduction

Object recognition plays an important role in intelligent system [1], such as aircraft recognition [2] and target tracking [3]. Before recognizing object, the interest region should be segmented by some effective algorithms of image segmentation such as Active Contour Models [4] and Level Set Method [5]. Then the recognition method recognize what the object in the interest region is. The SIFT [6] proposed by Lowe is designed for robust image feature detection. Shape Context [7] present a method to measure similarity between shapes. Shape Context is an effective approach. Then some methods based on shape context were proposed one after another, such as a visual shape descriptor using sectors and shape context of contour lines [8] and a modified shape context method for shape based object retrieval [9].

These above methods like [7] often have poor robustness on rotation invariance. Most of methods like [6] can hardly work well in binary images. This paper presents a novel approach, Superposition Proportion Method, which can overcome these difficulties. This approach need to make some prior models before matching, but it has lower computation complexity and much better robustness. As some good object

This work was supported by the National Natural Science Foundation of China under Grants No. 61571346, 61305040.

J. Pan et al. (eds.), *Intelligent Data Analysis and Applications*, Advances in Intelligent Systems and Computing 535, DOI 10.1007/978-3-319-48499-0_9

recognition methods are, this approach has robustness with respect to image scale, illumination, rotation and noise. In addition, this method can deal with occluded, shadowy and blurred images.

The proposed approach has three steps. Before recognizing what is the object, the image should be preprocessed. The preprocessing can eliminate disturbances. The next step is to recognize the regions contained by the object. This step can improve the robustness to disturbances of our method. Then the main direction should be calculated out. This main direction will provide rotation invariance for this method. Afterwards, rotate the model to make its main direction be parallel with it of the object, and then make the image be a minimum enclosing rectangle of the object. In addition the model image should be same with object image in size. Finally, calculate the superposition proportion.

The remainder of this paper is organized as follows. Section 2 introduces the new approach. Section 3 presents the experiments. Section 4 concludes the paper.

2 Superposition Proportion

The proposed approach define a main direction and use the main direction to obtain the rotation invariance. Regardless of the rotation, superposition proportion is closely related to the similarity, so the proposed method use the superposition proportion to estimate the similarity. Figure 1 depicts the flow chart of the proposed approach.

2.1 Eliminating Disturbances

The first task is to eliminate disturbances. Disturbances are ever changing, but the main region of object, which is always the biggest connected domain in foreground, are easily to recognize, so finding the main region is the most important step in this task. Because the proposed approach judges the object by inspecting overlap area in final step, the biggest connected domain in foreground should be the main part of object. If the biggest connected domain in foreground is not a part of object, this image is determined as an error input. In other words, the biggest connected domain in foreground is considered as the default main part of object.

Then it is necessary to remove disturbances which are usually much smaller than the main region of object or far away from the main region. Some assumed disturbance may be a small part of object actually, but it will not affect the calculation results as it is much smaller than the main region of the object.

$$S_{disturbances} = \{D \mid (distance(D, D_{main}) > d_{max}) \, or \, (area(D) < A_{min})\} \qquad (1)$$

where $S_{disturbances}$ is the set of disturbance, D is a connected domain, D_{main} is the main region of the object, $distance(D, D_{main})$ is the shortest distance between D and D_{main}, d_{max} is the threshold value os distance, $area(D)$ is the area of D and A_{min} is the threshold value of area.

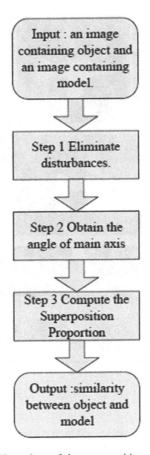

Fig. 1. Flow chart of the superposition proportion

Figure 2(a) is an image of an object which is a plane. Figure 2(b) is the original main region of the object. Figure 2(c) is the disturbances to the object. Figure 2(d) is a part of the main region of the object. Figure 2(e) is the main region of the object.

The following work is to enlarge the main region by making the connected domains, which are not disturbances, take parts in main region. Though as a result of the existence of interference these parts of main region are not connected, they are holistic.

2.2   Obtaining Main Direction

This step defines the main direction as the angle in that the variance of projection of the object is maximum. This is so important a step that can provide rotation invariance for the approach. Normally an object have only one main direction, so with obtaining main direction user can ignore rotation. In this step, it is necessary to use an optimization

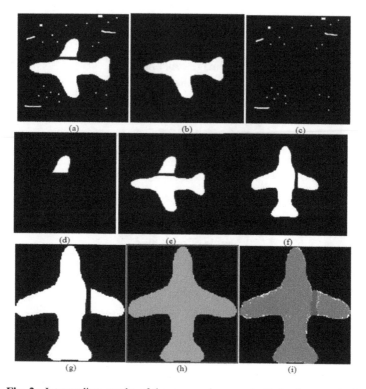

Fig. 2. Intermediate results of the proposed approach (Color figure online)

algorithm to compute the optimal angle, and then rotate the image to make the direction of optimal angle be vertical.

Projection is a vector, which equals the sum of an image towards one direction of α. Figure 2(f) presents the image which is processed by this step.

$$P_i = \sum_{x=i} I_{xy} \tag{2}$$

where P is a vector of projection, I_{xy} is the intensity of the pixel in (x,y) with the Y-axis being the direction of projection.

$$\begin{cases} \max \sigma = \sum_P (P_i - \overline{P})^2 \\ s.t. \alpha \in [0, \pi] \end{cases} \tag{3}$$

where σ is the variance of P; α is the angle of the direction in which the projection is obtained.

2.3 Computing the Superposition Proportion

The third step is to compute superposition proportion. This step demands the minimum enclosing rectangle of both model and object. It is important to make the model be same with the object in size to obtain the scale invariance. Finally similarity is obtained by computing the superposition proportion between object and model.

$$area_{max} = \max(area_{model}, area_{object})$$
$$area_s = area(image_{model} \,\&\, image_{object}) \tag{4}$$
$$similarity = area_s / area_{max}$$

where $area_{model}$ is the area of model; $area_{object}$ is the area of object; $area_s$ is the superposition of model and object.

Figure 2(g) and (h) are object and model respectively. The green area in Fig. 2(i) is $area_s$, the white area in Fig. 2(h) is $area_{max}$.

This step has a low computational complexity that can make the proposed approach run fast.

3 Experimental Results

This section shows the performance of Superposition Proportion by comparing with Shape Context [10]. Figure 3 is the first set of model and objects. The results of proposed method and Shape Context are shown in Table 1.

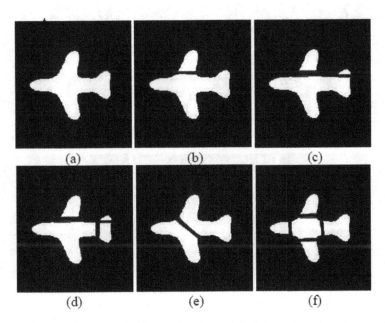

Fig. 3. The first set of model and objects. (a) is the model of plane. (b)–(f) are the objects those are occluded.

Table 1. The results of Fig. 4 in terms of occlusion.

Similarity between objects and model in Fig. 3	(b)	(c)	(d)	(e)	(f)
Shape Context	0.9075	0.9177	0.8133	0.6854	0.3045
Superposition proportion	0.9812	0, 9827	0, 9822	0, 9810	0, 9805

The results of Fig. 3(b) and (c) show that Shape Context have some robustness when disturbance is small. The results of Fig. 3(d)–(f) show that the bigger is the disturbance, the more poorly does the Shape Context perform. The performance of Superposition Proportion method has always been very good. The results show that Shape Context performs as badly poorly as object is occluded, but Superposition Proportion does very well no matter how poorly the object is occluded.

The Fig. 4 is the second set of model and objects, and the experimental results is presented in Table 2.

The results of Figs. 3(d) and 4(d) show that Superposition Proportion performs better than Shape Context. Tables 2 and 3 show that Superposition Proportion have perfect rotation invariance.

Figure 5 is the third set of model and objects. This set contains some similar objects and some dissimilar objects. Table 3 shows the overall performance of Superposition Proportion comparing with Shape Context.

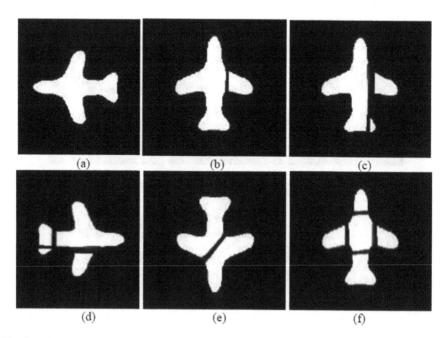

Fig. 4. The second set of model and objects. (a) is model of plane. (b)–(f) are objects those are different from the model in direction of main direction.

Table 2. The results of Fig. 4 in terms of rotation.

Similarity between objects and model in Fig. 4	(b)	(c)	(d)	(e)	(f)
Shape Context	0.9378	0.9176	0.2288	0.7890	0.3654
Superposition proportion	0.9812	0.9827	0.9822	0.9810	0.9805

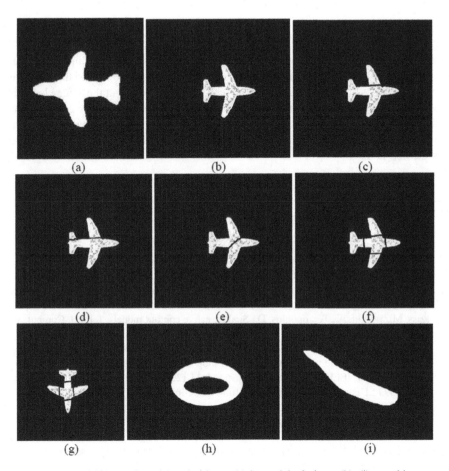

Fig. 5. The third set of model and objects. (a) is model of plane. (b)–(i) are objects.

Table 3. Results of Fig. 5 in terms of similarity.

Similarity	(b)	(c)	(d)	(e)
Shape Context	0.0267	0.0091	0.0082	0.0397
Superposition proportion	0.9531	0.9522	0.9488	0.9520
Similarity	(f)	(g)	(h)	(i)
Shape Context	0.0243	0.0092	0.9344	0.9409
Superposition proportion	0.9502	0.9502	0.3381	0.5015

It is easy to judge that Fig. 5(b)–(g) are similar with the model, and Fig. 5(h)–(i) are dissimilar with the model. The results of Fig. 5(b)–(g) show that Shape Context cannot recognize the similar objects. The results of Fig. 5(h) and (i) show that Shape Context can see the dissimilar objects as similar ones. The results of Fig. 5 show that the comprehensive performance of Superposition Proportion is excellent.

4   Conclusions

Overall, Superposition Proportion has good properties of rotation invariance and robustness to interference due to finding main region and direction. Also this proposed method is likeness friendly, meaning that it can recognize object in the same class with model. But this method may not works well in the images which have low discrimination between objects and backgrounds. This method will not performance well if the object is in a complex texture image. So the condition should be paid attention to when this method is used.

References

1. Wu, H.C.: Intelligent system. J.Inf. Sci. **103**(1–4), 135–159 (1997)
2. Li, Y., Chen, H., Mei, Y., et al.: Automatic aircraft object detection in aerial images. J. Proc. SPIE Int. Soc. Optical Eng. **5253**, 547–551 (2003)
3. Yilmaz, A., Shafique, K., Shah, M.: Target tracking in airborne forward looking infrared imagery. J. Image Vis. Comput. **21**(7), 623–635 (2003)
4. Kass, M., Witkin, A., Terzopoulos, D.: Snakes: active contour models. J. Int. J. Comput. Vis. **1**(4), 321–331 (1988)
5. Leventon, M.E., Grimson, W.E.L., Faugeras, O., et al.: Level set based segmentation with intensity and curvature priors. In: IEEE Workshop on Mathematical Methods in Biomedical Image Analysis. IEEE Computer Society, pp. 4–11 (2000)
6. Lowe, D.G.: Distinctive Image Features from Scale-Invariant Keypoints. J. Int. J. Comput. Vis. **60**(60), 91–110 (2004)
7. Belongie, S., Malik, J., Puzicha, J.: Shape matching and object recognition using shape contexts. J. IEEE Trans. Pattern Anal. Mach. Intell. **24**(4), 509–522 (2010)
8. Peng, S.H., Kim, D.H., Lee, S.L., et al.: A visual shape descriptor using sectors and shape context of contour lines. J. Inf. Sci. **180**(16), 2925–2939 (2010)
9. Madireddy, R.M., Gottumukkala, P.S.V., Murthy, P.D., et al.: A modified shape context method for shape based object retrieval. Springerplus **3**(1), 1–12 (2014)
10. Belongie, S., Malik, J., Puzicha, J.: Shape context: a new descriptor for shape matching and object recognition, pp. 831–837 (2000)

Piecewise Planar Region Matching for High-Resolution Aerial Video Tracking

Meng Yi[1,2(✉)] and Li-chun Sui[3]

[1] School of Electronic and Control Engineering, Chang'an University,
Xi'an 710064, China
yimeng0120@gmail.com
[2] Department of Civil, Environmental and Geodetic Engineering,
The Ohio State University, Colombus, OH 43210, USA
[3] School of Geological Engineering and Surveying, Chang'an University,
Xi'an 710064, China

Abstract. In order to tracking moving objects of aerial images, the frames and the scene is kept space consistency through image registration at the background. Due to high image resolution and large geographic deformation between different frames of aerial video, complicating the image registration. An piecewise planar region matching based image registration is introduced that can subdivide large frame into planar region, Image subdivision reduces the geographic distortions between aerial video, as it is usually the case of high-resolution aerial images. Then we can use select the most "useful" matching points that best satisfy the affine invariant space constraints are used to estimate the transformation model and register the images in a piecewise manner. Experiment result illustrate that the proposed method can register the high-resolution images and track the moving object in an aerial video.

Keywords: Aerial image registration · Piecewise planar · Affine invariant space constraints · Image tracking

1 Introduction

Aerial image registration is a common problem in many computer vision domains, including robot localization, panorama stitching, aerial mapping, ground moving target detection and tracking [1, 2]. In this study, our method is feature-based method, first identify the feature correspondences between image pairs and then utilize these corresponding control points to find transforms which register the image pairs.

Various methods for detecting control points in an image have been developed, Schmid et al. [3] have surveyed and compared various point detectors, finding the Harris detector [4] to be the most repeatable. Mikolajczyk [5] has proposed the scale-adapted Harris with automatic scale selection, However, this algorithm has the following two Shortcomings: First, this method computes the image gradient based on discrete pixel differences, and finite differences can provide a very poor approximation to a derivative. Second, there may be few feature points in the poor textural areas due to low contrast textures in local image regions. In this work, we resolve the above two

© Springer International Publishing AG 2017
J. Pan et al. (eds.), *Intelligent Data Analysis and Applications*, Advances in Intelligent Systems and Computing 535, DOI 10.1007/978-3-319-48499-0_10

shortcomings by applying optimal derivative filters based well-defined points [6]. We have the accurate location of the corner points by using the optimal derivative filters. Then the well-defined corners were found that do not move with slightly different focal lengths.

Feature matching is another important challenge in feature-based registration of very large multi-view images. In recent years, several feature matching methods have successfully applied in aerial image sequences registration, such as invariant descriptor [7] and invariant moments [8]. These methods work well when the sets of points to be matched are small, but hey become impractical or break down when images have local or nonlinear geometric differences. Methods to find correspondence between point sets with nonlinear geometric differences have been developed also [9, 10], but such methods are often slow and fail in the presence of outliers or sharp geometric differences between the images.

In this study, an accurate piecewise planar region matching based aerial image registration method is proposed. The strategy utilizes planar triangular of well-defined points and the most "useful" matching points to register the images in a piecewise method. Firstly, we use feature point based matching method generates a sparse correspondence between the two images, Here, well-defined points are the uniform-distributed feature points. Then the triangulation will be dynamically subdivided into sufficiently small regions by using Delaunay Triangulation, and A key aspect of our proposal is that of identifying the feature points that belong to the background using affine invariant space constraints, the corresponding background points are then used to estimate transformation model in each triangle region. Here, the corresponding background points in each region pair are in the same planar background with maximum reliability. Then two corresponding image triangular regions come from a planar patch of the scene, and they can be registered by the projective transformation. At last, we use Kalman filter [11] to filter jitter motion and obtain smooth object tracking results. This process will improve the registration accuracy and facilitate detection of moving objects.

2 Approach

2.1 Harris Interest Points Selection

The first step in image registration process is feature detection. Various point detectors in an image have been developed [12]. We use Harris corner detector [13] in our registration framework because of the abundance of corners in aerial images. Harris detector uses the second-moment matrix as the basis of its corner detections. For an image $I(x, y)$, the Harris detector based on the second-moment matrix can be expressed as:

$$M = \begin{bmatrix} G_{xt}^2 & G_{xt}G_{yt} \\ G_{yt}G_{xt} & G_{yt}^2 \end{bmatrix} * h \tag{1}$$

$$R = \det(M) - ktr^2(M) \tag{2}$$

Where h is the gaussian smoothing function, and G is the traditional image gradient.

The Harris detector provides good repeatability under rotation and various illuminations, unfortunately, it is sensitive to quantization noise and suffers from a loss in localization accuracy. In this section, instead of using the traditional image gradients, we use optimally first-order derivative filter with more accurate location and rotation-invariance [14]. Then we choose the best corners and remove those that are closer than a threshold distance of this corners. An example of feature points extracted is shown in Fig. 1.

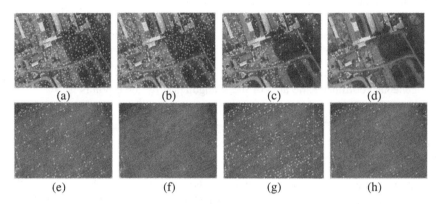

Fig. 1. Feature points detection results in different scene complexity. (a)–(e), (b)–(f) corners detected by Harris and optimal derivative filters based Harris when $\sigma = 1.5$ pixels. (c)–(g) corners detected by scene adaptation. (d)–(h)Stable corners detected by optimal derivative filters when $\sigma = 1$, $\sigma = 1.5$ and $\sigma = 2$ pixels.

2.2 Feature Matching

After feature points have been detected, the next step is to find the corresponding points between two point sets. Firstly, we construct the triangulation of each image. For a set of points $p_1, p_2, \ldots, p_n$, we compute the Delaunay Triangle (DT) by firstly computing its Voronoi Diagram(VD). The VD of a set of points is the division of a plane or space into regions for each point. The regions contain the part of the plane or space which is closer to that point than any other. With a given the VD, the DT is the straight line dual of the VD. A set of points are shown in Fig. 2(a), their VD is shown in Fig. 2(b) while their DT is shown in Fig. 2(c).

Then we use a hierarchical subdivision method to match the correspondence points and subdivide the triangulation, a more detailed description of each step as follows: (1) select delaunay graph edges in each image; (2) select two disjoint delaunay graph edges in each image and calculate the transformation function M. (3) count the number of other graph edges that also match with the existed transformation function M. (4) choose the transformation with the largest number of graph edges in the two images. (5) The hierarchical subdivision process (including the corners extraction, the

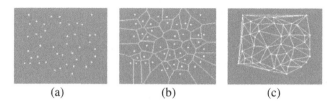

Fig. 2. (a) A set of points (b) Voronoi diagram (c) Delaunay triangulation.

matching propagation, and triangulation refining) is implemented iteratively until images are subdivided into small triangular corresponding triangles.

2.3 Affine Invariant Space Constraints

For three points $p_1 = (x_1, y_1, 1)$, $p_2 = (x_2, y_2, 1)$ and $p_3 = (x_3, y_3, 1)$ under the homogeneous coordinates, the area of the triangulation by three points is:

$$TAB(p_1, p_2, p_3) = \frac{1}{2} \begin{vmatrix} x_1 & y_1 & 1 \\ x_2 & y_1 & 1 \\ x_3 & y_1 & 1 \end{vmatrix} \tag{3}$$

Suppose the two matching point sets are $P' = (P_1, P_2, \cdots, P_m)$, $Q' = (Q_1, Q_2, \ldots, Q_m)$, then for any pair of matching points (P_i, Q_i), we can select k K-nearest neighbors (k = 5) in P' according to the Euclidean distance from P_i. we can select k K-nearest neighbors (k = 5) in P' according to the Euclidean distance from P_i. Where, m is the number of matching pairs, i is one matching pair, and the range of I is $1 \leq i \leq m$. We define that the nearest point to P_i is the starting point, and sort the set of points according to the counterclockwise, and define the sequence as $P_i = \{P_1, P_2, \ldots, P_k\}$, the matching point sequence Q' as $Q_i = \{Q_1, Q_2, \ldots, Q_k\}$.

Assuming $s_{xy} = |TAB(P_i, P_x, P_y)|$ represents a point (P_i, Q_i), $x = 1, 2, \ldots k$, $y = 1, 2, \ldots k$, $x \neq y$, we can get corresponding match point sets $s'_{xy} = |TAB(Q_i, Q_x, Q_y)|$. If (P_x, Q_x) and (P_y, Q_y) are correct matching point pairs, then:

$$\frac{s_{12}}{s'_{12}} \approx \frac{s_{23}}{s'_{23}} \approx \cdots \approx \frac{s_{k1}}{s'_{k1}} \tag{4}$$

Let $TAB_i = \{s_{12}/s'_{12}, s_{23}/s'_{23}, \ldots, s_{k1}/s'_{k1}\}$, we set the variance of $TAB_i = \{s_{12}/s'_{12}, s_{23}/s'_{23}, \ldots, s_{k1}/s'_{k1}\}$ as the affine invariant space constraint of point collection (P_i, Q_i), and let $AISC(P_i, Q_i) = Var(TAB_i)$. When $AISC(P_i, Q_i) \leq \xi$, (P_i, Q_i) is the correct one, where, ξ is the threshold.

The 3 unknown parameters of 2 N linear equations can be determined by substituting the N matching points into (4), and then we have $B = Am$, where

$$B = \begin{bmatrix} x_{t,1} - x_{t-1,1} \\ y_{t,1} - y_{t-1,1} \\ \vdots \\ x_{t,N} - x_{t-1,N} \\ y_{t,N} - y_{t-1,N} \end{bmatrix} \quad A = \begin{bmatrix} -y_{t-1,1} & 1 & 0 \\ x_{t-1,1} & 0 & 1 \\ & \vdots & \\ -y_{t-1,N} & 1 & 0 \\ x_{t-1,N} & 0 & 1 \end{bmatrix} \quad m = [\theta, \Delta X, \Delta Y]^{T} \quad (5)$$

After we obtained the relation between the camera and the scene, we can select Kalman filter [15] to filter jitter motion and obtain smooth object tracking results.

3   Experimental Results

The algorithm has been implemented in C ++ and all experiments have been carried out on 2.8 GHz Intel Core i7 computer with 16 GB 1600 MHz DDR3, with Windows 7 Professional Edition. Figure 3 shows 4 sets of the unmanned aerial vehicle(UAV) video sequences with size 320×240, including urban roads, rural area, fields farmland, Etc.

Figure 4 shows the reference image of a typical urban aerial image, which covers the better building areas. Figure 4(a) shows the same initial triangulation for the first layer, which constructed from 7 seed points. The numbers of matched points in the different stage are shown in Fig. 4(b)–(c).

(a)   (b)   (c)   (d)

Fig. 3. UAV video sequences.

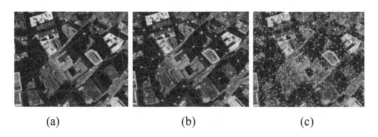

(a)   (b)   (c)

Fig. 4. (a) initial matching result, (b) further matching result, (c) final matching result.

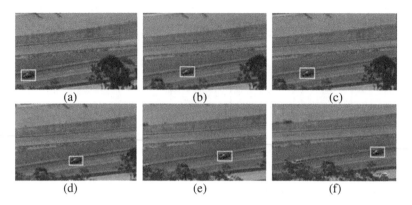

Fig. 5. City video tracking results.

The experimental results of car tracking in a city video images are shown in Fig. 5. We select six frames that start from the tenth frame, and each one was extracted every twenty frames. We can see from the figure that this method can track moving objects accurately. Figure 5 is the tracking result from the video images with occlusion.

4 Conclusions

In this paper, a piecewise planar region matching for high-resolution aerial video tracking was proposed. Keypoints are first extracted by optimized harris detector, then the correspondence between the feature points of the input image and reference image are found using RANSAC and hierarchical delaunay triangulation, then the input image with the correct mapping model is transformed and Kalman filter [15] is used to filter jitter motion and obtain smooth object tracking results. Experimental results demonstrate the proposed algorithm can achieve accuracy registration and detect slow moving targets on aerial video of urban and suburban scene.

5 Acknowledgment

This paper was supported by The Fundamental Research Funds for the Central Universities (310832151097) and China Postdoctoral Science Foundation Funded Project (2016M590912).

References

1. Zitova, B., Flusser, J.: Image registration methods: a survey. Image Vis. Comput. **21**(11), 977–1000 (2003)
2. Ali, S., Reilly, V., Shah, M.: Sneddon: motion and appearance contexts for tracking and reacquiring targets in aerial videos. In: IEEE CVPR, pp. 1–6 (2007)

3. Parag, T., Elgammal, A., Mittal, A.: A framework for feature selection for background subtraction. In: Proceedings of Computer Vision and Pattern Recognition, pp. 1916–1923 (2006)
4. Shum, H.-Y., Szeliski, R.: Construction of panoramic image mosaics with global and local alignment. Int. J. Comput. Vis. **36**(2), 101–130 (2000)
5. Lowe, D.G.: Distinctive image features from scale-invariant keypoints. Int. J. Comput. Vis. **15**(6), 415–434 (1997)
6. Mikolajczyk, K., Schmid, C.: A performance evaluation of local descriptors. IEEE Trans. Pattern Anal. Mach. Intell. **27**(10), 1615–1630 (2005)
7. Ke, Y., Sukthankar, R.: PCA-SIFT: a more distinctive representation for local image descriptors. In: Proceedings of IEEE International Conference on Computer Vision and Pattern Recognition, vol. 2, pp. 506–513 (2004)
8. Krish, K., Heinrich, S., Snyder, W.E.: Global registration of overlapping images using accumulative image features. Pattern Recogn. Lett. **31**(2), 112–118 (2010)
9. Rothwell, C.A., Zisserman, A.: Using projective invariants for constant time library indexing in model based vision. In: BMVC 1991 (1991)
10. Bay, H., Tuytelaars, T., Van Gool, L: SURF: speeded up robust features. In: Proceedings of the Ninth European Conference on Computer Vision, May 2008
11. Mundy, J.L., Heller, A.: Geometric Invariance in Computer Vision. MIT Press, Cambridge (1992)
12. Kahl, F., Heyden, A.: Using conic correspondences in two images to estimate the epipolar geometry. In: Proceedings of the International Conference on Computer Vision (1998)

Salient Object Detection Based on Histogram-Based Contrast and Guided Image Filtering

Pingping Zeng, Fanjie Meng$^{(\boxtimes)}$, Ruixia Shi, Dalong Shan, and Yanlong Wang

Institute of Intelligent Control and Image Engineering,
Xidian University, Xi'an 710071, China
ppzengxd@126.com, fjmeng@xidian.edu.cn,
{1067963479,243587539,951735431}@qq.com

Abstract. Detection of salient objects from images is gaining increasing research interest in recent years. Consider that the edge may not always be clear for some traditional saliency detection methods, a novel framework for saliency detection based on Histogram-based Contrast (HC) method and guided filter is proposed in this paper. Firstly, the algorithm based on HC is used to get a preliminary saliency map. Secondly, through a series of operation on the original input image, the approximate location of the salient object can be obtained. Then, the location map of salient object is generated by implementing guided image filtering on the approximate location and will be transformed into a binary image in the following process. Finally, the final saliency map can be achieved by fusing the binary map and the HC saliency map. The obtained saliency map not only inherits the good ability of HC method to maintain the internal information of the salient objects, but also has the advantage of the location map to well extract the edge of the salient object.

Keywords: Salient object detection · Histogram-based contrast · Guided image filtering · Saliency map

1 Introduction

People accept various kinds of information passively or actively in everyday life, it is necessary for us to extract useful and salient information quickly due to the fast pace of modern society. With the in-depth research, great deals of salient object detection algorithm have been proposed by researchers.

In the early time, Itti et al. proposed ITTI salient model [1] (also known as the ITTI model) based on the development of the Koch & Ullman model. The model mainly consists of three modules: characteristic extraction, saliency map generation and comprehensive saliency map generation. However, the algorithm is complex and has long running time. Hou [2] with some other people proposed a kind of saliency detection method of based on the SR (Spectral Residual), which get the spectral residual information in spectrum, and then construct the corresponding characteristic map in frequency domain. But while inhibiting the non-salient information, this method

© Springer International Publishing AG 2017
J. Pan et al. (eds.), *Intelligent Data Analysis and Applications*, Advances in Intelligent Systems and Computing 535, DOI 10.1007/978-3-319-48499-0_11

also inhibits the salient information, which makes the contrast of final saliency map is low. Shen [3] and others proposed a unified model of low rank matrix restoration (LMRM), which combines the traditional low level features with a higher level of instruction to detect salient targets. But the problem is that the feature extracted is not accurate enough so that the saliency map is not so satisfactory. Starting from the frequency domain, Achanta et al. [2] propose a frequency tuned (FT) method that directly defines pixel saliency using a pixel's color difference from the average image color. The elegant approach, however, only considers first order average color, which can be insufficient to analyze complex variations common in natural images. Cheng Mingming [5] proposed a method based on histogram-based contrast (HC) to detect saliency. The HC saliency map allocates the saliency value according to the difference of color between pixels, thus, producing a saliency map with full resolution. It is called histogram-based contrast method, because it processes efficiently by using histogram-based method, and uses the smooth operation of color space to control the defects of quantification. As an improvement over HC-maps, Cheng et al. incorporate spatial relations to produce region-based contrast (RC) maps. The saliency value of a region is calculated using a global contrast (GC) score, measured by the region's contrast and spatial distances to other regions in the image. The disadvantage of the algorithms Cheng proposed is that they are not good at texture background suppression.

This paper proposes a saliency detection algorithm based on HC and image guided filter [6]. Image guided filters can be used to smooth the image while retaining the edge information of the image. Image guided filters are used in many image processing applications, such as image smoothing, image enhancement, HDR compression, image matting and joint sampling. The method proposed well preserves the edge information of salient objects by edge detection. At the same time, compared with other existing methods, it performs better in restraining the non-salient background information, which makes foreground of salient object more prominent.

2 Salient Object Detection

The proposed object detection framework consists of two major components: extraction of original salient object and location of salient object. The final saliency map is generated by fusing these two parts using the fusion principle based on regional energy. The flowchart of the proposed algorithm is shown in Fig. 1.

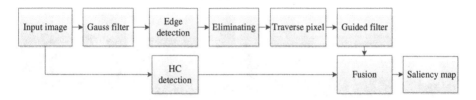

Fig. 1. Flow chart of the algorithm proposed in this paper

2.1   Extraction of Original Salient Object

The method based on HC is proposed by Cheng et al., which is based on the color statistical characteristics of the input image. It is a method that calculates the saliency of the image according to the characteristic difference between the pixels, that is, the saliency value of a pixel is defined by the color contrast between itself and other pixels in the image. The method uses the LAB color characteristics of the three channels simultaneously to calculate the pixel difference. LAB color characteristic is the most commonly used image characteristic, as it is a kind of color space which is closest to human vision. In this paper, the input image is the RGB image I shown in Fig. 2, and the saliency of pixel I_k is defined in formula (1) [5]:

$$S(I_k) = \sum_{\forall I_i \in I} D(I_k, I_i) \tag{1}$$

Where $D(I_k, I_i)$ is the color distance measurement in LAB space for pixel I_k and pixel I_k. The formula above can be carried out in accordance with the order of pixels:

$$S(I_k) = D(I_k, I_1) + D(I_k, I_2) + \cdots + D(I_k, I_N) \tag{2}$$

N is the number of pixels in the image. Through the formula, it can be seen that due to the neglect of the spatial relationship, the pixels that have the same color value get the same saliency value in this definition. Therefore, the above formula can be realigned to merge. We merge the pixels that have the same value c_i. So get the saliency value of each color. The whole process can be expressed by the following formula (3):

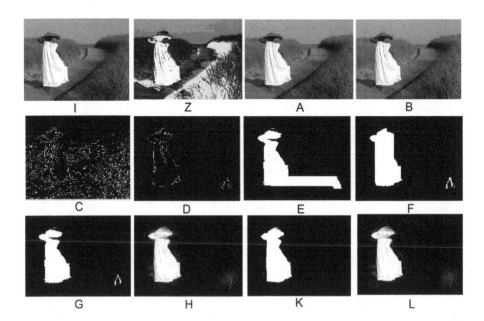

Fig. 2. Location of the salient object

$$S(I_k) = S(c_m) = \sum_{i=1}^{n} f_i D(c_m, c_i) \tag{3}$$

Where c_m is the color value of pixel I_k n is the sum of the color number that image contains. f_i is the probability of c_i to appear in the image I. The original saliency map generated by this method is image Z shown in Fig. 2. It can be seen from the figure that the salient object detection method based on histogram contrast can well expresses the internal information of salient object. But its ability to suppress the background information of non-salient background information is not so good, which leads to the introduction of lots of background.

2.2 Location of Salient Object

Pretreatment. Convert the original image I to the gray image A. Conduct Gaussian filtering [7] on it and filter out the high frequency part of the input image in order to reduce the background noise of the original image. At the same time, it also weakens the edge of background. In this paper, the input image of Gaussian filter is A as shown in Fig. 2, of which the filtering function is selected as the Gauss function [8] according to the human visual characteristics:

$$G(x, y) = \frac{1}{2\pi\sigma^2} exp\left(-\frac{x^2 + y^2}{2\sigma^2}\right) \tag{4}$$

Among the formula, G (x, y) is a circular symmetric function [9]. the smoothing function of which can be controlled. We can get a smooth image by convolving image with input image. The formula is as following:

$$F(x, y) = G(x, y) * A(x, y) \tag{5}$$

In this paper, the smooth image obtained by Gauss filtering is B in Fig. 2.

Log Operator for Edge Detection. Log (Laplacian of Gaussian) algorithm [10] is a method of edge detection proposed by Marr and Hildreth according to the human visual characteristics. LOG operator first use Gaussian low pass filter to smooth the image, and Laplacian operator is used to find the steep of the image. Finally it uses zero gray value to get binary closed and connected profile, eliminating all interior points. The advantage of this method is that both the image is smoothed and the noise is reduced, and the isolated noise points and smaller structures are filtered out. In this paper, the image obtained from the edge detection of B is image C shown in Fig. 2.

Eliminate Edge Lines with Fewer Pixels. The edge of salient object of image can be extracted by Log operator edge detection, but at the same time there remain a number of free edges of background. Legacy background edge is very small compared to the edge profile of salient objects in the image. So THIS design an algorithm to compute all pixels in the edge line of image C as N_1, $N_2 \cdots N_n$. Record the number of pixels in the edge line that contains the maximum number of pixels as N_{max}, and set up threshold as $\alpha \cdot N_{max}$. When the number of pixels the edge profile contains satisfies the formula $N_i < \alpha \cdot N_{max}$, then eliminate this edge profile. After many experiments, the results show that the best eliminating effect can be obtained when a = 0.5. In this way, we get the edge contour of the salient object roughly. The results generated by this step are shown in Fig. 2 as image D. We can see from the result that this method is a good way to eliminate most of background information contained in image C.

Initial Location of Salient Object Of the image D which is gotten by last step, we traverse the image pixels from top to bottom, left to right. First each line of image D is traversed, and record the coordinates of the i_{th} line and j_{th} column pixel as X (i, j). Then record the coordinates of the pixel of value 1 first encountered in i_{th} line as X (i, 1), and the coordinates of value 1 pixel in last column as X (i, 2). Set the value of the pixels between X (i, 1) and X (i, 2) as 1, then get a binary image E as shown in Fig. 2. Then traverse pixels in each column, record the coordinate of the pixel in the p_{th} line and q_{th} column as Y (p, q). In the q_{th} column, record the coordinate of the pixel of value 1 the first line encountered as Y(1,q), the last line encountered as Y(2,q). Change the value of the pixels between Y(1,q) and Y(2,q) in the q_{th} column into 1 to get image F as shown in Fig. 2. Do and operation on image E and G, then make thresholding processing to the results to get a binary image G as shown in Fig. 2. The white area of image G is salient region of the input image while the black area is non-salient background area. Through the processing of this step, we get the initial saliency map with a distinct profile of the salient object and achieve the initial location of the salient object.

Guided Filtering. The guided filtering of the image is linear variable filter process. The guidance image is required according to the specific application in advance, but also can be directly taken as the input image. In this paper the guidance image is original image I, the input image is binary image G; the output image after guided filtering is image H as Fig. 2 shows. For the ith pixel of the output image, the calculation method can be mathematically formalized as (6) [6]:

$$h_i = \sum_j W_{ij}(I)G_j \tag{6}$$

Where i and j are pixel labels, W_{ij} is the filter kernel function [6]:

$$W_{ij}(I) = \frac{1}{|\omega|^2} \sum_{k:(i,j) \in \omega_k} \left(1 + \frac{(I_i - \mu_k)(I_J - \mu_k)}{\sigma_k^2 + \varepsilon} \right) \tag{7}$$

Where W_k represents the kth kernel function window, $|\omega|$ is the number of pixels within the window; μ_k and σ_k^2 is the mean and variance value of guidance image I within the window is smooth prior. Compared with the conventional filtering kernel function, the guidance kernel function has good performance of edge preserving and detail enhancing. The theoretical analysis of its performance is described in detail in the literature [6]. The image H we obtained through guided filtering well maintains the edge details and suppresses the non-salient background information.

2.3   The Final Saliency Map

Preliminary saliency map Z and an initial location map K of the salient object are obtained from above steps. In order to obtain a final map with better visual characteristics and more details, the fusion of initial saliency map and location map is divided into two parts based on regional energy principle [11]. Take the neighborhood window of the two images as 3×3, and calculate the area energy $E_K(i,j)$, $E_Z(i,j)$ respectively, which represents the energy of the area centered by pixel (i,j):

$$E_K(i,j) = \sum_{i=1}^{3} \sum_{j=1}^{3} [T_K(i,j)]^2 \tag{8}$$

$$E_Z(i,j) = \sum_{i=1}^{3} \sum_{j=1}^{3} [T_Z(i,j)]^2 \tag{9}$$

Where $T(i,j)$ represents the 3×3 neighborhood window who is centered at (i,j). Then, corresponding to the pixel in location map, if the value of pixel $P_K(i,j)$ is 1, then we set the value of corresponding pixels in final saliency map according to the bigger the better principle. As shown in formula (10):

$$P(i,j) = \begin{cases} P_K(i,j) & E_K(i,j) \geq E_Z(i,j) \\ P_Z(i,j) & E_K(i,j) \leq E_Z(i,j) \end{cases} \tag{10}$$

On the contrary, if the value of pixel $P_K(i,j)$ is 0, then we set the value of corresponding pixels in final saliency map according to the smaller the better principle. As shown in formula (11):

$$P(i,j) = \begin{cases} P_K(i,j) & E_K(i,j) \geq E_Z(i,j) \\ P_Z(i,j) & E_K(i,j) \leq E_Z(i,j) \end{cases} \tag{11}$$

Through comparing the area energy in preliminary saliency map Z and an initial location map K, and selecting the pixel with higher energy in the salient area, the pixels with lower energy in background, we get the fusion image as final saliency map L shown in Fig. 2. From the resulted map L, it can be seen that the method proposed not only inhibits the non-salient background, but also well maintains the advantage of HC algorithm of well extracting the internal information of salient object.

3   The Result of Experiment and Analysis

This chapter optionally chooses 9 test images from the image database as shown in Fig. 3. Then we do test on them with the method proposed by this paper, and compare them with 6 kinds of most frequently-used visual attention model. The 6 methods includes: (1) ITTI model [1]; (2) SR model [2]; (3) FT model [4]; (4) HC model [5]; (5) GC model [5]; (6) LMRM model [3]. In this paper, the 6 methods along with the

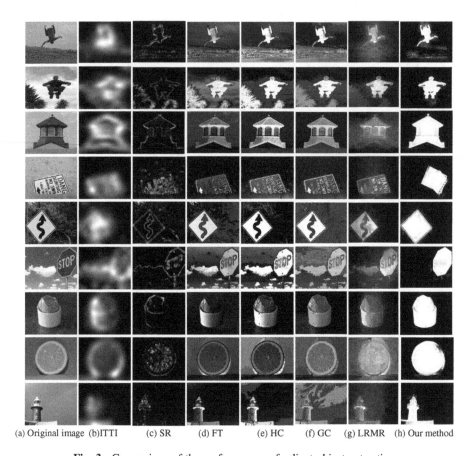

(a) Original image (b)ITTI   (c) SR   (d) FT   (e) HC   (f) GC   (g) LRMR (h) Our method

Fig. 3. Comparison of the performances of salient object extraction

proposed algorithm are used to test the same image in the same scene. The experimental results are shown in Fig. 3.

Compared with other 6 existing algorithms, it can be seen that not only does the saliency detection algorithm proposed by this paper extract the salient objects more clearly, but it also well maintains the edge of salient objects as well as effectively inhibits non-salient background information. So from the perspective of visual effects, our method is better than other algorithms.

In order to evaluate the effectiveness of the proposed algorithm accurately and comprehensively, this paper also adopt objective evaluation criteria of Mean Precision and Recall rate as well as the comprehensive evaluation index F-Measure to evaluate the proposed algorithm. The specific data is shown in Table 1.

As the results show, both the recall rate and the F-Measure index of the proposed algorithm are the highest among the seven methods, which further proves the superiority of the proposed algorithm.

Table 1. Mean Precision, Recall and F-Measure indexes of test images in Fig. 3

Algorithm	Precision	Recall	F-Measure
ITTI	0.6012	0.6121	0.6066
SR	0.5323	0.3450	0.4186
FT	0.7402	0.5499	0.6310
HC	0.5857	0.6064	0.5959
GC	0.8502	0.5513	0.6689
LRMR	0.7716	0.6776	0.7216
Our method	0.8181	0.9901	0.8959

4 Conclusion

Based on the combination of the histogram contrast and the guide filter, a new salient object detection algorithm is proposed in this paper. The main idea is to locate the salient object and suppress non-salient background information in an image through combined use of the edge detection method and the guide filter. Then, based on the fusion criterion of region energy, the initial saliency map obtained by the HC method is fused with the initial saliency map. The algorithm proposed in this paper overcomes the disadvantage of the HC method that sometimes it is not good enough for the background suppression, and retains the advantage that it can detect the internal information of the salient object very well. The results shows that the salient object detection method based on the histogram contrast and the guided filter can accurately detect the salient target. In addition, comparisons with six state-of-the-art algorithms also demonstrate the superiority of the algorithm proposed in this paper.

Acknowledgments. This work was supported by the National Natural Science Foundation of China under Grant 61305040 and the Fundamental Research Funds for the Central Universities under Grant JB161305.

References

1. Itti, L., Koch, C., Niebur, E.: A model of saliency-based visual attention for rapid scene analysis. IEEE Trans. Pattern Anal. Mach. Intell. **20**(11), 1254–1259 (1998)
2. Hou, X.D., Zhang, L.Q.: Saliency detection: a spectral residual approach. In: IEEE Conference on Computer Vision and Pattern Recognition, pp. 1–8 (2007)
3. Shen, X.H., Wu, Y.: A unified approach to salient object detection via low rank matrix recovery. In: IEEE Conference on Computer Vision and Pattern Recognition, pp. 853–860 (2012)
4. Achanta, R., Hemami, S., Estrada, F., et al.: Frequency-tuned salient region detection. In: IEEE Conference on Computer Vision and Pattern Recognition, pp. 1597–1604 (2009)
5. Cheng, M.M., Zhang, G.X., Mitra, N.J., et al.: Global contrast based salient region detection. In: Proceedings of 2011 IEEE Conference on Computer Vision and Pattern Recognitions, pp. 409–416 (2011)

6. He, K., Sun, J., Tang, X.: Guided image filtering. In: Daniilidis, K., Maragos, P., Paragios, N. (eds.) ECCV 2010, Part I. LNCS, vol. 6311, pp. 1–14. Springer, Heidelberg (2010)

7. Kovesi, P.: Fast almost-gaussian filtering. In: International Conference on Digital Image Computing: Techniques and Applications, pp. 121–125. IEEE Computer Society (2010)

8. Zhang, J., Lian, Y., Dong, L., et al.: A new method of fuzzy edge detection based on Gauss function. In: 2010 The 2nd International Conference on Computer and Automation Engineering (ICCAE), pp. 559–562. IEEE (2010)

9. Katz, A.J.: Generalized hebbian· learning for principal component analysis and automatic target recognition, systems and method: US, US 6894639 B1 (2005)

10. Zhao, F., Desilva, C.J.S.: Use of the Laplacian of Gaussian operator in prostate ultrasound image processing. In: Proceedings of the International Conference of the IEEE Engineering in Medicine and Biology Society, vol. 2, pp. 812–815 (1998)

11. Xing, S.X., Chen, T.H., Li, J.X.: Image fusion based on regional energy and standard deviation. In: International Conference on Signal Processing Systems, pp. V1-739–V1-743. IEEE (2010)

A New Method for Extraction of Residential Areas from Multispectral Satellite Imagery

Rui Xu[1,2], Yanfang Zeng[3(✉)], and Quan Liang[1,2]

[1] College of Information Science and Engineering, Fujian University
of Technology, No.3, Xueyuan Road, University Town, Fuzhou 350118, China
[2] Fujian Provincial Key Laboratory of Big Data Mining and Applications,
Fujian University of Technology, No.3, Xueyuan Road,
University Town, Fuzhou 350118, China
[3] College of Tourism, Fujian Normal University,
No.1, Keji Road, University Town, Fuzhou 350117, China
yanfang99084@163.com

Abstract. This paper presents a new method for residential areas extraction from multispectral satellite imagery. First, the image is preprocessed so that non-residential areas noises such as vegetation regions, shadows and water areas are removed. Second, the preprocessed image is classified into two categories by texture features: residential areas class and non-residential areas class. Finally, the residential areas class is refined by using a series of morphology operations. The experimental results show that this method is effective in extracting residential areas from multispectral satellite imagery.

Keywords: Residential areas · Classification · Gabor filter · Multispectral satellite imagery

1 Introduction

Residential areas extraction from satellite images has become an active and important research area in recent decades. The extracted results have diverse important applications, including city planning, disaster assessment, and environmental change research, etc. The availability of accurate, high-resolution multispectral images delivered by the new generation of sensors has provided the potential to discriminate very subtle details in urban scenes. These images provide us with rich spatial and spectral information and can be utilized for residential areas extraction.

Various approaches have been developed to address the residential areas extraction issue. Most researchers utilize classification to extract residential areas from remote sensing images [1–3]. Fauvel et al. [4] classified the residential areas by considering the fusion of multiple classifiers. First, data are processed by each classifier separately and the algorithms provide for each pixel membership degrees for the considered classes. Second, a fuzzy decision rule is used to aggregate the results provided by the algorithms according to the classifiers' capabilities. Benediktsson et al. [5] used morphological and neural approaches to extract urban areas. First, the composition of geodesic opening and closing operations of different sizes is used in order to build a differential

© Springer International Publishing AG 2017
J. Pan et al. (eds.), *Intelligent Data Analysis and Applications*, Advances in Intelligent Systems and Computing 535, DOI 10.1007/978-3-319-48499-0_12

morphological profile that records image structural information. Second, feature extraction and feature selection is applied. Third, a neural network is used to classify the features from the second step.

However, it is difficult to extract residential areas accurately by using classification alone due to the complexity of remote sensing imagery. To tackle this problem, we present a new multistage method for extracting residential areas. In this method, we firstly remove vegetation regions, shadows and water areas by the Normalized Difference Vegetation Index (NDVI) and the Hue-Saturation-Intensity (HSI) Index. Then the preprocessed image is classified into residential areas class and non-residential areas class by texture features. Next, a set of morphology operations are used to refine the residential areas. Experiments show that the proposed method is able to achieve a good performance in residential areas extraction.

The remainder of this paper is organized as follows. The proposed approach is presented in Sect. 2. Experimental results are presented in Sect. 3. Conclusions are provided in Sect. 4.

2   Methodology

The proposed method aims to improve the accuracy of residential areas extraction from multispectral satellite imagery. The workflow of the proposed method is presented in this section. The organization of this method is shown in Fig. 1. The details of each step are introduced as follows.

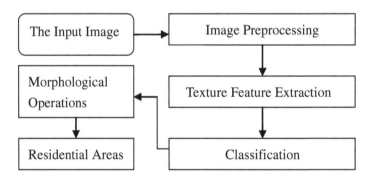

Fig. 1. Flowchart of the proposed method.

2.1   Image Preprocessing

To improve the effectiveness of our method, image is preprocessed to remove noises such as vegetation, shadows, and water areas.

Firstly, we calculate the vegetation regions based on the NDVI which is obtained from the Red and Near-Infrared (NIR) by the following formula [6]

$$NDVI = (NIR - Red)/(NIR + Red) \tag{1}$$

Next, a HSI color space is converted from a Near-Infrared-Red-Green false color space. Then shadow regions and water areas are detected by the HSI index which is calculated by the following formula

$$HSI\ index = (Saturation - Intensity)/(Saturation + Intensity) \tag{2}$$

Finally the noises are removed by Otsu's threshold [7].

2.2 Gabor Filter Texture Feature Extraction

The Gabor filters [8] have received considerable attention, because the characteristics of certain cells in the visual cortex of some mammals can be approximated by these filters. In addition, these filters have been shown to posses optimal localization properties in both spatial and frequency domain, and thus are well suited for texture classification problems.

2D Gabor function [9] is

$$g(x,y) = \left(\frac{1}{2\pi\sigma_x\sigma_y}\right)\exp\left[-\frac{1}{2}\left(\frac{x^2}{\sigma_x^2} + \frac{y^2}{\sigma_y^2}\right) + 2\pi jWx\right] \tag{3}$$

where, W is a Gaussian function of complex modulation frequency. Its Fourier transform can be shown as

$$G(u,v) = \exp\left\{-\frac{1}{2}\left[\frac{(u - W)^2}{\sigma_u^2} + \frac{v^2}{\sigma_v^2}\right]\right\} \tag{4}$$

where $\sigma_\mu = 1/2\pi\sigma_x$, $\sigma_v = 1/2\pi\sigma_y$. And σ_x, σ_y characterize the spatial extent and frequency bandwidth of the Gabor filter. Let $g(x,y)$ be the mother Gabor wavelet, then this self similar set of Gabor functions $g_{m,n}(x,y)$ can be obtained by rotating and scaling $g(x, y)$ through the generating function

$$g_{mn}(x,y) = a^{-m}g(x',y'), a > 1, m,n \in ingeters \tag{5}$$

where $x' = a^{-m} (xcos\theta + ysin\theta)$, $y' = a^{-m} (-xsin\theta + ycos\theta)$, $\theta = n\pi/K$, K represents the total number of direction ($n\in [0, K]$), by changing the value of m and n, we can obtain a set of directions and different scales of the filter.

For a given image $I (x, y)$, its Gabor wavelet transform can be defined as

$$W_{mn}(x,y) = \iint I(x,y)g_{mn}^*(x - x_1, y - y_1)dx_1dy_1 \tag{6}$$

where g_{mn}^* represents the complex conjugate. The mean and standard deviation of the magnitude of the orientation bands, which are used to construct the texture feature vector, can be calculated as

$$\mu_{mn} = \iint |W_{mn}(x,y)|d_x d_y \tag{7}$$

$$\sigma_{mn} = \sqrt{\iint (|w_{mn}(x,y)| - \mu_{mn}|)^2 d_x d_y} \tag{8}$$

In this paper, Gabor filter is applied to extract texture feature of the image. After determining m scales and n orientations of Gabor filter, the mean μ_{mn} and standard deviation σ_{mn} can be calculated. The texture feature vector is defined as follows

$$F = [\mu_{00}\sigma_{00}, \mu_{01}\sigma_{01}, \ldots \ldots, \mu_{m-1,n-1}\sigma_{m-1,n-1}] \tag{9}$$

2.3   Residential Areas Extraction

As SVM is a supervised classification method which is as good as or significantly better than other competing methods in most cases [10], we employ SVM to improve the robustness and the accuracy of residential areas extraction. The classification method is presented in this section. The details of each step are listed as follows.

Step1: The image is preprocessed by the method mentioned in Subsect. 2.1, the vegetation, shadows, and water areas are removed.
Step2: The texture features are extracted as the input features.
Step3: For each class, 10 % of ground truth data are selected as the training samples.
Step4: Use Gaussian kernel, and select the parameters, namely, penalty parameter C and kernel parameter, to train the SVM classifier.
Step5: Classify the whole image using the classifier trained in step 4.
Step6: Assign 1 to residential areas class, and assign 0 to others.

The algorithm of classification is able to remove most false residential regions. However, misclassified regions still exist, and further processing is necessary to improve the reliability of residential areas extraction. In general, residential areas have these characteristics: (1) Residential areas do not have small areas; regions with small areas can be regarded as noisy and should be removed, (2) Residential areas are massive, continuous areas, with no holes;The residential areas are refined by using a series of morphology operations such as corrosion and open operation.

3   Experiment

In this experiment, the study area is a part of Fuzhou City image which was recorded by the QuickBird optical sensor. The study area has a spatial dimension of 800*800 pixels. The spatial resolution is 2.44 m per pixel. Figure 2(a) shows the study area of this experiment. We chose the SVM algorithm in the supervised classification, because the training data are interpreted in polygons, and the SVM can adopt this kind of training data. Figure 2(b)–(c) show the result of classification using spectral feature and the result of classification using GLCM (Grey Level Co-occurrence Matrices) [11] texture feature. Figure 2(d) shows the result of the proposed method. Compared with the results showed in Fig. 2(b) and (c), the proposed method can distinguish residential areas more effectively. To quantify the performance of the proposed method, three accuracy measures are used to evaluate it. These measures are: (1) *completeness = TP/ (TP + FN)*; (2) *correctness = TP/(TP + FP)*; and (3) *quality = TP/(TP + FP + FN)*. The variables *TP*, *FN*, and *FP* denote true positive, false negative, and false positive, respectively. In this experiment, the proposed method gives the completeness of 90.09 %, the correctness of 91.91 %, and the quality of 83.47 %. From the measures result, it can be seen that the proposed method shows a good performance in the extraction of residential areas from multispectral satellite imagery. The superposition result of the original image and extracted residential areas was shown in Fig. 2(f).

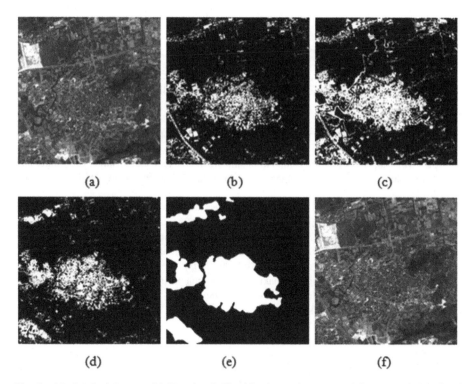

Fig. 2. (a) Original image. (b) Result of Classification using spectral feature. (c) Result of Classification using GLCM texture feature. (d) Result of the proposed method. (e) Result of filtering by a series of morphology operations. (f) The superposition result of the original image and the extracted residential areas by the proposed method.

4   Conclusion

This paper presents a method to extract residential areas, and this method focuses on extracting large-scale residential areas. The experimental result shows that the proposed method is effective. There are two advantages about our method described as follows: Firstly, the preprocessed method is effective in reducing noises by removing vegetation regions, shadows and water areas. Secondly, the designed classifier can effectively distinguish residential areas. However, the proposed method is semi-automatic which requires manual intervention. Future work will therefore focus on improving the degree of automation.

Acknowledgments. This work is supported by the project of science and technology of Fujian province (2016H0001), the project of Fuzhou Municipal Science and Technology Bureau (2015-G-53), and the key project of science and technology of Fujian province (2014H6006).

References

1. Vatsavai, R.R., Bright E., Varun C., et al.: Machine learning approaches for high-resolution urban land cover classification: a comparative study. In: International Conference and Exhibition on Computing for Geospatial Research & Application, pp. 1–10, Washington, DC, USA (2011)
2. Tao, C., Tan, Y., Zou, Z.R., et al.: Unsupervised detection of built-up areas from multiple high-resolution remote sensing images. IEEE Geosci. Remote Sens. Lett. **10**(6), 1300–1304 (2013)
3. Chen, H., Tao, C., Zou, Z.R., et al.: Automatic urban area extraction using a gabor filter and high-resolution remote Sensing imagery. Geomat. Inf. Sci. Wuhan Univ. **38**(9), 1063–1067 (2013)
4. Fauvel, M., Chanussot, J., Benediktsson, J.A.: Decision fusion for the classification of urban remote sensing images. IEEE Trans. Geosci. Remote Sens. **44**(10), 2828–2838 (2006)
5. Benediktsson, J.A., Pesaresi, M., Amason, K.: Classification and feature extraction for remote sensing images from urban areas based on morphological transformations. IEEE Trans. Geosci. Remote Sens. **41**(9), 1940–1949 (2003)
6. Holben, B.N.: Characteristics of maximum-value composite images from temporal avhrr data. Int. J. Remote Sens. **7**(11), 1417–1434 (1986)
7. Ohtsu, N.: A Threshold selection method from gray-level histograms. IEEE Trans. Syst. Man Cybern **9**(1), 62–66 (1979)
8. Fogel, I., Sagi, D.: Gabor filters as texture discriminator. Biol. Cybern. **61**(2), 103–113 (1989)
9. Manjunath, B.S., Ma, W.Y.: Texture features for browsing and retrieval of image data. IEEE Trans. Pattern Analy. Mach. Intell. **18**(8), 837–842 (1996)
10. Burges, C.J.C.: A tutorial on support vector machines for pattern recognition. Data Min. Knowl. Discov. **2**(2), 121–167 (1998)
11. Haralick, R.M., Shanmugam, K., Dinstein, I.: Texture features for image classification. IEEE Trans. Syst. Man Cybern. **3**(6), 610–621 (1973)

Tooth Segmentation from Cone Beam Computed Tomography Images Using the Identified Root Canal and Harmonic Fields

Shi-Jian Liu[1,2,3], Zheng Zou[4(✉)], Ye Liang[5], and Jeng-Shyang Pan[1,2]

[1] Shcool of Information Science and Engineering, Fujian University of Technology,
No. 3 Xueyuan Road, University Town, Minhou, Fuzhou 350118, Fujian, China
{liusj2003,jengshyangpan}@fjut.edu.cn
[2] Key Laboratory of Big Data Mining and Applications of Fujian Province,
No. 3 Xueyuan Road, University Town, Minhou, Fuzhou 350118, Fujian, China
[3] The Key Laboratory for Automotive Electronics and
Electric Drive of Fujian Province, No. 3 Xueyuan Road, University Town,
Minhou, Fuzhou 350118, Fujian, China
[4] Shcool of Information Science and Engineering, Central South University,
No. 932 South Lushan Road, Changsha 410083, Hunan, China
zouzheng84@csu.edu.cn
[5] Department of Stomatology, Xiangya Hospital of Central South University,
No. 100 Xiangya Road, Changsha 410005, Hunan, China

Abstract. In this paper, a novel method is introduced to segment tooth from Cone Beam Computed Tomography images. Different from traditional methods, the root canal centerline identified by graph theory based energy minimization problem is applied as prior knowledge aiding for the segmentation. Besides, though we use the idea of contour tracking strategy as adopted by most published methods based on slice-by-slice basis, within a slice, the segmentation is based on the harmonic field theory, which makes our method superior to the traditional ones. Effect and efficiency of ours are proved by the experiments.

Keywords: Segmentation · Root canal · Cone Beam CT · Harmonic field

1 Introduction

There are two major ways nowadays to capture in vivo dental information into its digital format. The first one is to use a laser scanner, which is a popular three dimensional (3D) approach and can be used to capture the geometry of crown surface accurately. However, as we know, the laser beams cannot reach the

This work is partially supported by the Doctoral Innovation Fund of Hunan (CX2012B066), the Scientific Research Project in Fujian University of Technology (GY-Z160066). The authors also gratefully acknowledge the helpful comments and suggestions of the reviewers, which have improved the presentation.

J. Pan et al. (eds.), *Intelligent Data Analysis and Applications*, Advances in Intelligent Systems and Computing 535, DOI 10.1007/978-3-319-48499-0_13

internal of object, therefore, it is unable to capture the root of tooth. As another option, Cone Beam Computed Tomography (CBCT) techniques are widely used in the clinic, since the X-ray it adopted can pass through human body and tell the differences among different organs. Actually, the CBCT images play an very important role in the computer-aided diagnosis, treatment planning and virtual surgery for dental routines such as root canal therapy, where information of the root is crucial to their successes. In order to be used by the computer, a primary step of the CBCT based approach is to distinguish tooth from the rests, which is the aim of this work. In other words, we want to find a feasible way to segment individual teeth from CBCT images. The segmentation of tooth from CBCT images is challenging according to Gao et al. [11]. Firstly, the tooth is very close to surrounding objects, such as adjacent teeth (see Fig. 1(a)) or jaw bones (seen Fig. 1(c)). Therefore, many suspicious boundaries exist and they even disappeared where adjacent teeth touch each other. Secondly, the tooth contour also suffers from topological change among different images. For example, the first molar in Fig. 1(b) can be seen as one connected region, while becomes three region in Fig. 1(c). Thirdly, the resolution of tooth is limited, because the image represents the entire head and some surroundings. As Fig. 1(d) demonstrated, the region of tooth in red rectangle is relatively small comparing to the entire region of image in blue rectangle. Fourthly, the CBCT images may be very noisy for the reason of reduced radiation dose, which is for the consideration of healthy. Lastly, efficiency is important due to large number of slices for each tooth.

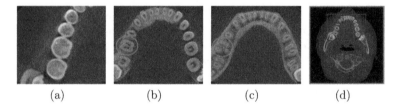

| (a) | (b) | (c) | (d) |

Fig. 1. Challenges of the tooth segmentation lives in: adjacent teeth (a), topological changes (b, c), and limited resolution (d).

1.1 Related Work

Lots of methods have been proposed for medical image segmentation. Some of them are relatively simple, such as threshold based method and region growing [17,19]. Others may based on more some complex methodologies or procedures, such as max-flow/min-cut based graph cut [5] method and contour evaluation based Snakes [2] or Level Set [8] algorithm. But none of them can be used as general approach suitable for ever organ segmentation problems, which is the conclusion of a survey [7].

Since the CT images can be view as an entire 3D volume or a sequence of 2D images, the existed tooth segmentation methods can be classified into two categories, namely the volume based and slice based approaches. 3D mesh models (or well know as template) which represents the mean shape and prior knowledge of each teeth are generally required for the segmentation and reconstruction of volume based methods [4,6]. In order to get the desire models, training processes are commonly needed based on some previous constructed databases, which make the method of this kind extremely complex and time costing. By contrast, slice based methods treat CT images in their very native way. Namely, they work on a slice-by-slice basis, and in order to track the segmentation boundary (or contour) in consecutive slices, the contour of segmented slices will be offered as prior knowledge for the segmentation of adjacent slices generally. Among the slice based methods, Level Set is the most adopted algorithm [10–13,16]. The segmentation of a slice is achieved by process named contour evolution, it is a iterative procedure initialized by a given contour, and evolved under the target of minimization of a specified energy function. Therefore, the choice of the initial contour and energy function are critical to the results. Parameter tunings are generally required in order to get a excellent result for different cases [9]. Liao et al. [14,15] adopt the thin-plate splines to connect control points specified by user interactions as the segmentation boundary of tooth in CT volume. Though user interactions can offer high level knowledge aiding for the segmentation, which are good especially for the Ground Truth generation, but too many of them will be time-costing and lead to tedious user experiences.

This paper also adopt the slice-by-slice basis, but different from the existed method, the centerline of root canal will be identified in the first place and used as prior knowledge in additional to the contour of adjacent slices. Furthermore, harmonic field theory other than algorithms such as level set is used in this paper to find the segmentation boundary. The rest of this paper will be organized as follows. In Sect. 2, a root canal centerline identification method based on graph theory is introduced. Section 3 presents the harmonic field based segmentation method. Then, Sect. 4 shows some experimental results and assessments. Finally, the Sect. 5 concludes the paper.

2   Graph Theory Based Root Canal Centerline Identification

Instead of solid object, within a tooth there exist some spaces anatomically where soft tissues, such as the blood vessels and nerve, can be found. Therefore, in the CBCT images, low intensity areas appear inside the tooth in contrast with the high intensity bone tissues as can be seen in Fig. 1.

A root canal is the tubular structure within the root of a tooth. In this paper, we treat it as an important prior knowledge based on the fact that a tooth can be located as long as its root canals are identified. In order to achieve that, centerline, which is an efficient descriptive way of root canal, will be firstly defined as a path formed by consecutive edges between two selected terminal points in the graph constructed from CT volume. Then, similar to the level set approaches, energy terms will be special designed and assigned to every edges of the graph on the condition that the centerline will possess the minimum energy values than any other path's between the two specified terminals. Finally, the classic shortest weighted path algorithm can be used to solve the minimization problem where energy term of each edge is the weight.

2.1   Definition of the Graph and Centerline

A 3D volume can be seen as the superposition of two-dimensional (2D) CT images in the order of their Z-coordinations. The 2D pixel in images is referred to as voxel in 3D volume. A most intuitive while inefficient way of graph construction is to use every voxels as graph vertexes and the line determined by each two neighboring voxels as graph edges, because the scale of the graph will become too complicated to be processed generally. An improvement is to use a subset of entire voxels in the construction.

Actually, a skeletonization method [1] is used to remove the voxels, which definitely no belonging to the root canal region, during the graph construction. And we use the classic Dijkstra algorithm for the shortest weighted path problem solving.

A discrete version of the root canal centerline is a path of graph we constructed, which can be described by a sequence of graph vertexes $\{x_1, x_2, ..., x_k\}$, where x_1 and x_k are two vertexes closest to the terminal points of centerline. If we assign each edge a energy term $E(x_i, x_{i+1}), (1 \leq i \leq k-1)$, than the energy of the path can be depicted as Eq. 1.

$$\sum_{1}^{k-1} E(x_i, x_{i+1}) \tag{1}$$

2.2   Designation of the Energy

In this paper, two aspects are considered in the designation of energy term as described above, including the position $P(x_i, x_{i+1})$ and tangential direction $D(x_i, x_{i+1})$ of edge x_i, x_{i+1} as described by Eq. 2.

$$E(x_i, x_{i+1}) = \alpha P(x_i, x_{i+1}) + (1-\alpha)D(x_i, x_{i+1}) \tag{2}$$

Volume intensities and structure tensor are adopted to formulate the $P(x_i, x_{i+1})$ and $D(x_i, x_{i+1})$ for the consideration that the centerline is supposed to exist in a low intensity tubular structure. And the α is set to 0.5 to balance them. In additional, every voxel intensities of input volume V is firstly reversed V' according to Eq. 3 so as to enhance the root canal region.

$$I'(x) = M - I(x) \tag{3}$$

where $M, I(x)$ denotes the maximum intensity of volume V and the intensity at voxel x respectively.

Then, the proposed $P(x_i, x_{i+1})$ and $D(x_i, x_{i+1})$ will meet the Eqs. 4 and 5.

$$P(x_i, x_{i+1}) = \frac{2M - (I(x_i) + I(x_{i+1}))}{2(M - m)} \tag{4}$$

$$D(x_i, x_{i+1}) = 1 - \frac{\sqrt{\lambda_2{}^2\lambda_3{}^2(v \bullet e_1)^2 + \lambda_1{}^2\lambda_3{}^2(v \bullet e_2)^2 + \lambda_1{}^2\lambda_2{}^2(v \bullet e_3)^2}}{\lambda_1\lambda_2} \tag{5}$$

where $M, m, I(x_i)$ are the maximum, minimum intensity and the intensity at voxel x_i of the modified volume V'. $\lambda_i, e_i(i = 1, 2, 3)$ are eigenvalue and eigenvector of structure tensor of x_i subjected to $\lambda_1 \geq \lambda_2 \geq \lambda_3$, v is the vector from x_i to x_{i+1}, $\bullet$ denotes the dot product operation.

3   Harmonic Fields Based Tooth Segmentation

Our segmentation strategy is to construct a scalar field φ (i.e. the harmonic field) of image pixels, which satisfies $\Delta\varphi = 0$ (Δ is the Laplacian operator) and subject to particular Dirichlet boundary conditions (namely the constrains). More details of the harmonic field can be found in our previous work [20].

3.1   Segmentation Within a Slice

If we view the Laplacian operator of each pixel together as a matrix L (i.e., the Laplacian matrix), the element L_{ij} of L can be expressed as Eq. 6.

$$L_{ij} = \begin{cases} \sum_k W_{ik}, & if \ i = j, k \neq i \\ -W_{ij}, & if \ i \ are \ j \ are \ neighbors \\ 0, & otherwise \end{cases} \tag{6}$$

where W_{ij} is the weighting function should be designed for different cases. Let $I(i), I(j), L_{ij}$ denote the intensity of pixel i, j and the distance between them, then our W_{ij} meets the Eq. 7.

$$\omega_{ij} = e^{-\frac{(I(i)-I(j))^2}{L_{ij}{}^2}} \tag{7}$$

In order to solve the linear system $L\varphi = 0$, the constrains should be specified. For example, pixels of two categories are preset to 0 and 1 respectively, which are the minimum and maximum scalar of the harmonic field. Generally, one category includes pixels which belonging to the aims of segmentation (e.g., the tooth in our case), others are background pixels. Therefore, pixels belonging to the identified root canal centerlines and image boundaries can be treated as the two sets of constrains respectively.

After acquiring the harmonic field by solving the linear system determined by Laplacian matrix and the constrains, one of the harmonic iso-lines can be used as the segmentation boundary.

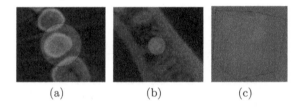

Fig. 2. Assignment of the S_c (a), S_t (b), and VOI (c) by user interactions.

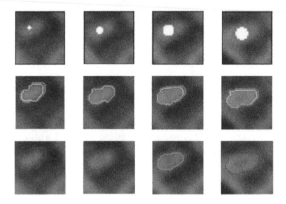

Fig. 3. Demonstration of contour tracking. The first row depicts the constrains, second row shows the harmonic field and its iso-lines, the third row presents the acquired contours.

3.2 The Contour Tracking Procedure

We name the entire segmentation procedure as contour tracking because the harmonic field based segmentation method described above is carried out iteratively from S_c to S_t to track the tooth contours and solve the problem. S_c and S_t are two extreme image slices closest to the crown side and tip side of tooth which include root canal centerline. User interactions are involved to assign the S_c and S_t (see Fig. 2(a) and (b)), as well as the Volume of Interest (VOI, see Fig. 2(c)) which includes the objective tooth.

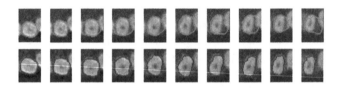

Fig. 4. Comparison results between the MS (the first row) and ours (the second row).

During the iteration, the shape of acquired contour in current slice will guide for the extraction of contour in the next slice. Figure 3 demonstrated the entire procedure.

4   Experiments and Result

The proposed system is implemented on a personal computer (2.67 GHz Core Duo processor and 4 GB memory) based on the Visualization Toolkit [18] and using C++ for programming. The CBCT images are provided by Xiangya Hospital of Central South University.

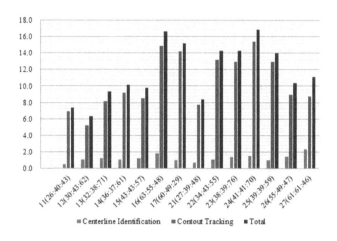

Fig. 5. Average time consumption (in seconds).

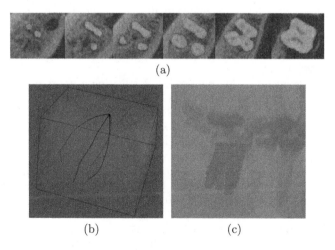

Fig. 6. Demonstrate of the segmentation results. (a) Segmentation contours of some slices. (b) The identified centerlines. (c) Overlapping of the final result and the volume.

Some experiments are carried out to evaluate the proposed method. Firstly, a Morphological Snakes (MS) algorithm [3] is adopted for the comparison. As showed in Fig. 4, our method is more accurate than the MS.

Then, the average time consumption for root canal centerline identification and contour tracking also are recorded (see Fig. 4) grouped by teeth indexes (i.e., from 11 to 17 and 21 to 27 according to the Federation Dentaire Internationale notation method).

Finally, the segmentation result of a tooth with multiple root canal is depicted in Fig. 6 to demonstrate the effect of the proposed method.

5   Conclusion and Future Works

This paper presents a novel segmentation method of tooth from CBCT images. Firstly, the root canal centerline identified by graph theory based energy minimization problem is applied as prior knowledge aiding for the segmentation. Besides, though we use the idea of contour tracking strategy as adopted by most published methods, within a slice, the segmentation is based on the harmonic field theory, which makes our method superior to the traditional ones. Effect and efficiency of ours are proved by the experiments.

As can be seen in the experiment, the major time consumption of our method is the cost for harmonic field generation. Though strategies for accelerating such as VOI extraction have been achieved in this work, further improvement is still a direction of our future works.

References

1. Abeysinghe, S.S., Baker, M., Chiu, W., Ju, T.: Segmentation-free skeletonization of grayscale volumes for shape understanding. In: IEEE International Conference on Shape Modeling and Applications, pp. 63–71, June 2008
2. Alvarez, L., Baumela, L., Henriquez, P., Marquez-Neila, P.: Morphological snakes. In: 2010 IEEE Conference on Computer Vision and Pattern Recognition, pp. 2197–2202, June 2010
3. Alvarez, L., Baumela, L., Marquez-Neila, P., Henriquez, P.: A real time morphological snakes algorithm. Image Process. Online 2, 1–7 (2012)
4. Barone, S., Paoli, A., Razionale, A.V., Savignano, R.: 3D reconstruction of individual tooth shapes by integrating dental cad templates and patient-specific anatomy. In: 2014 International Design Engineering Technical Conferences and Computers and Information in Engineering Conference, pp. 79–87 (2014)
5. Boykov, Y., Jolly, M.-P.: Interactive organ segmentation using graph cuts. In: Delp, S.L., DiGoia, A.M., Jaramaz, B. (eds.) MICCAI 2000. LNCS, vol. 1935, pp. 276–286. Springer, Heidelberg (2000). doi:10.1007/978-3-540-40899-4_28
6. Buchaillard, S.I., Ong, S., Payan, Y., Foong, K.: 3D statistical models for tooth surface reconstruction. Comput. Biol. Med. 37(10), 1461–1471 (2007)
7. Bulu, H., Alpkocak, A.: Comparison of 3D segmentation algorithms for medical imaging. In: Proceedings of the 26th IEEE International Symposium on Computer-Based Medical Systems, pp. 269–274 (2007)

8. Chung, G., Vese, L.A.: Image segmentation using a multilayer level-set approach. Comput. Vis. Sci. **12**(6), 267–285 (2009)
9. Derraz, F., Beladgham, M., Khelif, M.: Application of active contour models in medical image segmentation. In: International Conference on Information Technology: Coding and Computing, vol. 2, pp. 675–681, April 2004
10. Gao, H., Chae, O.: Touching tooth segmentation from CT image sequences using coupled level set method. In: 5th International Conference on Visual Information Engineering, pp. 382–387, July 2008
11. Gao, H., Chae, O.: Individual tooth segmentation from CT images using level set method with shape and intensity prior. Pattern Recogn. **43**(7), 2406–2417 (2010)
12. Hosntalab, M., Aghaeizadeh Zoroofi, R., Abbaspour Tehrani-Fard, A., Shirani, G.: Segmentation of teeth in CT volumetric dataset by panoramic projection and variational level set. Int. J. Comput. Assist. Radiol. Surg. **3**(3), 257–265 (2008)
13. Ji, D.X., Ong, S.H., Foong, K.W.C.: A level-set based approach for anterior teeth segmentation in cone beam computed tomography images. Comput. Biol. Med. **50**, 116–128 (2014)
14. Liao, S.-H., Tong, R.-F., Dong, J.-X.: 3D whole tooth model from CT volume using thin-plate splines. In: International Conference on Computer Supported Cooperative Work in Design, vol. 1, pp. 600–604 (2005)
15. Liao, S.-H., Han, W., Tong, R.-F., Dong, J.-X.: A hybrid approach to extracting tooth models from CT volumes. In: Martin, R., Bez, H., Sabin, M. (eds.) IMA 2005. LNCS, vol. 3604, pp. 308–317. Springer, Heidelberg (2005). doi:10.1007/11537908_18
16. Pavaloiu, I.B., Vasilateanu, A., Goga, N., Marin, I., Ilie, C., Ungar, A., Patracu, I.: 3D dental reconstruction from CBCT data. In: 2014 International Symposium on Fundamentals of Electrical Engineering, pp. 1–6, November 2014
17. Pohle, R., Toennies, K.D.: Segmentation of medical images using adaptive region growing (2001)
18. Schroeder, W., Martin, K.M., Lorensen, W.E.: The Visualization Toolkit, An Object-Oriented Approach To 3D Graphics (2006)
19. Verma, B., Sardana, H.K.: Real time image segmentation using watershed algorithm on FPGA. Int. J. Eng. Sci. Technol. **3**(6), 4518–4522 (2011)
20. Zou, B.J., Liu, S.J., Liao, S.H., Ding, X., Liang, Y.: Interactive tooth partition of dental mesh base on tooth-target harmonic field. Comput. Biol. Med. **56**(2015), 132–144 (2015)

Thresholding Method Based on the Relative Homogeneity Between the Classes

Hong Zhang[1($\boxtimes$)] and Wenyu Hu[2]

[1] School of Automation, Xi'an University of Posts and Telecommunications,
Xi'an 710121, Shaanxi, China
zhmlsa@xupt.edu.cn
[2] Fujian Provincial Key Laboratory of Big Data Mining and Applications,
Fujian University of Technology, Fuzhou, China

Abstract. Image binarization method focusing on the objects emphasized much the homogeneity of the object gray level distribution, can overcome some shortcomings of famous Otsu's method. In this paper, the specific information of image is considered, the gray and neighborhood average histogram are applied, a more detailed description of the threshold calculation is presented. A new threshold discriminant criterion is proposed with relative homogeneity information both of foreground and background. Using valley point neighborhood histogram information, a modified discriminant analysis is constructed. The experimental results on images with a minimal distribution difference between classes show that, compared to both Otsu's and focusing on the objects methods, the proposed method has not only better segmentation accuracy, but also better adaptability for images.

Keywords: Image segmentation · Otsu's method · Threshold discriminant · Gray histogram · Relative homogeneity

1 Introduction

Image segmentation is one of the most fundamental and important tasks in image analysis [1–4]. In [5], the authors described the 40 thresholding applications methods, which are categorized into the six groups based on the type of information used.

In earlier research, Otsu [6] proposed maximum between-class variance criteria to select the best threshold, which is regarded as one of the classic techniques and clustering criterion [5]. Nonetheless, the Otsu's method will provides a biased threshold when the gray level distribution functions have either unequal variances. As an attempt to overcome the inherent defect of Otsu's method, Hou [7] proposed minimum class variance thresholding (MCVT) method. Chen [8] analyzed the limitations of Otsu's criteria and developed a new binarization method by defining a

H. Zhang—This work is supported by the National Science Foundation of China (No. 61340040), the Provincial Natural Science Foundation research project of Shanxi (No. 2012JQ8045), and the Provincial Education project of Shaanxi (No. 15JK1682).

© Springer International Publishing AG 2017
J. Pan et al. (eds.), *Intelligent Data Analysis and Applications*, Advances in Intelligent Systems and Computing 535, DOI 10.1007/978-3-319-48499-0_14

discriminant criterion. Extensive research [9–12] has been already conducted to introduce new robust thresholding techniques based on the class variance thresholding segmentation.

In this paper, based on Chen's [8] method, a more detailed description of the threshold calculation is presented using image gray distribution histogram information. In addition, we analyzed the shortcoming of Chen's method, developed a new thresholding discriminant with relative homogeneity between foreground and background. With the specific histogram distribution images, a valley point neighborhood histogram information was introduced to modify the discriminant criterion. The experiments results on real images demonstrate that the approaches can perform not only visually better segmentation but can better adapt to the images of different gray distribution characteristics.

2 Image Binarization Focusing on Objects

2.1 Otsu's Method

Assume the gray levels of the given image ranges in $\{0, 1, 2, \cdots, L-1\}$, where L is the total number of gray levels of the image, $M \times N$ is the size. The gray value of pixel (x, y) is expressed by i, $f(i)$ is the total pixels number of gray level i, then the probability of gray level i is expressed as:

$$p(i) = \frac{f(i)}{M \times N}, \; i = 0, 1, \cdots, L-1$$

Otsu's binarizing method selects an optimal threshold t for a given image by a within-class discriminant function [6]:

$$\sigma_W^2(t) = P_1(t)\sigma_1^2(t) + P_2(t)\sigma_2^2(t)$$
$$= \sum_{i=0}^{t-1} p(i)(i - \mu_1(t))^2 + \sum_{i=t}^{L-1} p(i)(i - \mu_2(t))^2 \qquad (1)$$

Where, $\sigma_1^2(t) = \sum_{i=0}^{t-1} (i - \mu_1(t))^2 \frac{p(i)}{P_1(t)}$, $\sigma_2^2(t) = \sum_{i=t}^{L-1} (i - \mu_2(t))^2 \frac{p(i)}{P_2(t)}$. $P_1(t) = \sum_{i=0}^{t} p(i)$,

$P_2(t) = \sum_{i=t+1}^{L-1} p(i)$ is the prior probability, $\mu_1(t) = \frac{\sum_{i=0}^{t-1} ip(i)}{P_1(t)}$, $\mu_2(t) = \frac{\sum_{i=t+1}^{L-1} ip(i)}{P_2(t)}$ is the mean

of two classes. And thus, the optimal threshold value t^* is:

$$t^* = Arg \max_{0 \leq t \leq L-1} \sigma_W^2(t) \qquad (2)$$

2.2   Binarization Focusing on Objects

From above, in nature, Otsu's method views both the object and background as having uniformity or homogeneity of gray levels. However, such a case just holds partially. For some images, the pixels of object may have more uniformity or homogeneity in gray level distribution than the background, it means that the background possesses more likely a heterogeneous and non-uniform distribution, and naturally produces many different and diverse gray levels. Therefore, a biased threshold estimate will possibly be resulted with adopting a single mean to represent the background. In addition, Otsu's criterion only takes the gray levels into account but neglects their spatial distribution and contextual relationship, it neglects the continuity of the object gray levels and thus more likely over-stresses a role played by the numbers of the two class pixels rather than the gray levels themselves.

In order to remedy such shortcomings, Chen [8] defined an alternative discriminant criterion, it focus primarily on the object to be segmented and at the same time, only assumes that object has gray level homogeneity. As a result, the threshold selection can still be considered as a classification problem, an optimal gray level t* is selected which makes the following new criterion $J_{LC}(t)$ minimized:

$$J_{LC}(t) = \left(\frac{P_1(t)}{P_2(t)}\right)^\alpha \frac{\sum\limits_{(x,y)\in o}\left[\lambda(g(x,y) - m)^2 + (1 - \lambda)(\bar{g}(x,y) - m)^2\right]}{\sum\limits_{(x,y)\notin o}\left[\lambda(g(x,y) - m)^2 + (1 - \lambda)(\bar{g}(x,y) - m)^2\right]} \qquad (3)$$

Then, the optimal threshold t^* is:

$$t^* = \arg\min_{0 \le t \le L-1} J_{LC}(t) \qquad (4)$$

Here, $m = \frac{1}{|O|}\sum\limits_{(x,y)\in O} g(x,y)$, $P_1(t) = \frac{1}{|O|}$, $P_2(t) = \frac{1}{N-|O|}$, O is the set of pixels belonging to the object, (x, y) denotes the gray level of the pixel at (x, y), $\bar{g}(x, y)$ is the neighboring average, W is a window centered at (x, y) and $|W|$ is usually taken as 3×3 or 5×5, m is the mean of gray levels of the object foreground, N is the total number of pixels, $\alpha(\alpha \ge 0)$ is an exponent and adjusts $(P_1(t)/P_2(t))^\alpha$ to achieve some trade-off and λ ($0 \le \lambda \le 1$) also is an adjustable parameter to trade off the proportion. In formula (4), the numerator only measures the object-class similarity or scatter degree, in which the second term reflects local spatial continuity or homogeneity. The more similar the pixels in the object, the smaller the scatter and thus the smaller the value of the numerator is. And the denominator in (3) measures the background-class dissimilarity to the object. A larger value of the denominator implies that two classes are better separated even when background is heterogeneous.

This criterion more focuses on both the similarity of the object class itself and the dissimilarity of the background to the object, for better avoiding the problem probably incurred by the heterogeneity of background. Even if the background is not consistent, the object can be better separated. It seems somewhat consistent with a human intuition of segmenting image: an object with a uniform distribution in a diverse and non-uniform scene can easier be discerned and identified by our human eyes.

3 Application Based on Image Histogram Information

As described in Sect. 2, the threshold discriminant gives a new criterion for threshold selection. Based on this criterion, the threshold value is selected focusing on the homogeneity object, and local neighborhood mean is introduced. It is not only better to segmentation but also has a relatively obvious removal ability for noise.

In the above discriminant, it is implied that m can represent all, and t can also represent all. In the selection of image threshold, if we take into account the difference of the gray level histogram and the neighborhood average histogram, a more detailed description can be given.

Here, based on the Otsu's within-class variance method, we introduced the image gray histogram and the pixel neighborhood average information. The variables in the formula (4) can be described in detail, then thresholding criterion is obtained, i.e.

$$J_O(t) = \left(\frac{P_1(t)}{P_2(t)}\right)^{\alpha} \frac{\sum_{i \in O}\left[\lambda \cdot p(i)(i - \mu(t))^2 + (1 - \lambda) \cdot p(i)(i - \bar{\mu}(t))^2\right]}{\sum_{i \notin O}\left[\lambda \cdot p(i)(i - \mu(t))^2 + (1 - \lambda) \cdot p(i)(i - \bar{\mu}(t))^2\right]} \tag{5}$$

Where, $P_1(t) = 1/\sum_{i \in O} p(i)$, $P_2(t) = 1/\sum_{i \notin O} p(i)$ are the prior probability of object and background, $\mu(t)$ denotes the object area mean value of original image, $\bar{\mu}(t)$ denotes the object area mean value of neighborhood average image. $\mu(t) = P_1(t) \sum_{i \in O} ip(i)$, λ, α are adjusting parameters.

Figure 1 is the segmented results of Color image by using formula (5). Figure 1(a) is original image, Fig. 1(b) is the segmented result by $J_O(t)$ (denoted as 1d_O), Fig. 1 (c) and Fig. 1(d) are the segmented results by 1d and 2d Otsu's methods, (denoted as 1d_Otsu and 2d_Otsu) respectively. From Fig. 1, we can see that the segmentation results can't be obtained using 1d_Otsu and 2d_Otsu methods, while the object extraction result by $J_O(t)$ is complete and clear, as shown in Fig. 1(b). The parameters α and λ were calculated by 0.5 according to the paper [8].

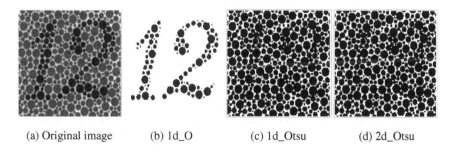

(a) Original image   (b) 1d_O   (c) 1d_Otsu   (d) 2d_Otsu

Fig. 1. The thresholded results of color image

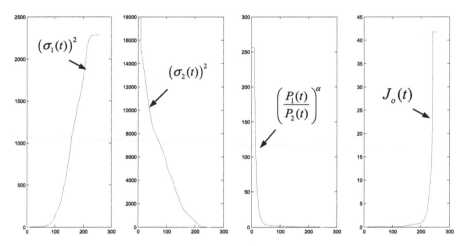

Fig. 2. Experimental data curve of rice image

However, in the application of some images, we found the limitation of this thresholding criterion. To simplify, assuming formula (5) is expressed as:

$$J_O(t) = \left(\frac{P_1(t)}{P_2(t)}\right)^\alpha \frac{(\sigma_1(t))^2}{(\sigma_2(t))^2} \tag{6}$$

In   which, $(\sigma_1(t))^2 = \sum_{i \in O} \left[\lambda \cdot p(i)(i - \mu(t))^2 + (1 - \lambda) \cdot p(i)(i - \bar{\mu}(t))^2\right]$,

$(\sigma_2(t))^2 = \sum_{i \notin O} \left[\lambda \cdot p(i)(i - \mu(t))^2 + (1 - \lambda) \cdot p(i)(i - \bar{\mu}(t))^2\right]$

As an example, the experimental data of Rice image are described, as shown in Fig. 2. With the increase of the threshold t, $(\sigma_1(t))^2$ increases monotonically, while $(\sigma_2(t))^2$ decreases monotonically about t. This result produces an meaningless extreme value. Naturally, the applicability of discriminant function is limited.

Analyzing the reason, the composition of $J_O(t)$ only includes the degree of consistency of object (background) with respect to the other pixels, the less the pixels, the better the relative uniformity. That is, with increasing of pixels number, the relative uniformity will be more and more poor, $J_O(t)$ shows monotonous.

4   Thresholding Method Based on the Relative Homogeneity Between Classes

From above section we know, the optimal threshold value can not be obtained only by considering the homogeneity of object (background). In thresholding, the pixels with good homogeneity may be located in the low value region, also be located in the high value region, therefore, the above discriminant can be described as the following:

$$J_{O1}(t) = \left(\frac{P_1(t)}{P_2(t)}\right)^\alpha \frac{\sum_{i=0}^{t-1}\left[\lambda \cdot p(i) \cdot (i - \mu_1(t))^2 + (1 - \lambda) \cdot p(i) \cdot (i - \overline{\mu_1}(t))^2\right]}{\sum_{j=t}^{L-1}\left[\lambda \cdot p(j) \cdot (j - \mu_1(t))^2 + (1 - \lambda) \cdot p(j) \cdot (j - \overline{\mu_1}(t))^2\right]} \qquad (7)$$

$$J_{O2}(t) = \left(\frac{P_2(t)}{P_1(t)}\right)^\alpha \frac{\sum_{i=t}^{L-1}\left[\lambda \cdot p(i) \cdot (i - \mu_2(t))^2 + (1 - \lambda) \cdot p(i) \cdot (i - \overline{\mu_2}(t))^2\right]}{\sum_{j=0}^{t}\left[\lambda \cdot p(j) \cdot (j - \mu_2(t))^2 + (1 - \lambda) \cdot p(j) \cdot (j - \overline{\mu_2}(t))^2\right]} \qquad (8)$$

Where, $P_1(t) = 1 / \sum_{i=0}^{t-1} p(i)$, $P_2(t) = 1 / \sum_{i=t}^{L-1} p(i)$, $\mu_1(t) = \sum_{i=0}^{t-1} ip(i) / \sum_{i=0}^{t-1} p(i)$, $\mu_2(t) = \sum_{i=t}^{L-1} ip(i) / \sum_{i=t}^{L-1} p(i)$.

$J_{O1}(t)$ denotes the criterion function in which homogeneity is better in low gray level region, $J_{O2}(t)$ denotes the criterion that homogeneity is better in high level region. For an image, the information will be lost if only using formula (7) or (8), for the regional internal uniform information, we can combine the two formulas and construct a new threshold formula as follows:

$$J_{OB}(t) = J_{O1}(t) + J_{O2}(t) \qquad (9)$$

Then, the optimal thresholding value is:

$$t^* = \arg \min_{0 \leq t \leq L-1} J_{OB}(t) \qquad (10)$$

In the above formula, it is not needed to distinguish the object is in low or high gray level region, which takes into account the uniformity and the integrity of the two classes. Especially for some images, the object region is homogeneous, and the homogeneity of background is better either.

Comparing formula (9) with within-class Otsu formula (1), we can see the numerator are the same, while the denominator is used to measure the relative degree of uniformity. For the two classes, although there is greater uniformity difference, but compared to the other class, there is a better homogeneity characteristics. Therefore, we defined this method as the thresholding method based on the relative homogeneity within-classes.

For demonstrating the effectiveness, we use the three images shown in Figs. 3 and 4 to compare segmenting effects with $J_{O1}(t)$, $J_{O2}(t)$, $J_{OB}(t)$ and 1d _Otsu.

Figures 3 and 4(a) are original images, Figs. 3 and 4(b) are the segmented result by $J_{O1}(t)$ (1d_O1), Figs. 3 and 4(c) are the segmented result by $J_{O2}(t)$ (1d_O2), Figs. 3 and 4(d) are the segmented results by $J_{OB}(t)$ (1d_OB), Figs. 3 and 4(e) are the segmented results by 1d_Otsu. The thresholding value are shown in Table 1. Comparing with the segmented results of the four methods, we can see that the proposed relative homogeneity method can obtain the most complete object extraction results.

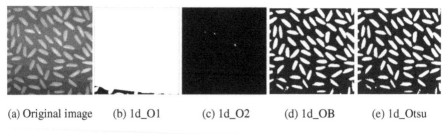

(a) Original image   (b) 1d_O1   (c) 1d_O2   (d) 1d_OB   (e) 1d_Otsu

Fig. 3. The thresholdeds result of rice image

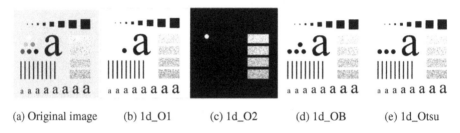

(a) Original image   (b) 1d_O1   (c) 1d_O2   (d) 1d_OB   (e) 1d_Otsu

Fig. 4. The thresholded results of letter image

5 Modified Thresholding Method Based on the Relative Homogeneity Between Classes

In Sect. 4, using the relative homogeneity method between the object and background, the rationality of threshold selection can be improved, and the better segmentation effect can be obtained. However, for some images with complex background or specific distributions, the applicability is still limited, as shown in Figs. 5 and 6(b) of the Aerial and Cont images. Although, using $J_{OB}(t)$ method, the monotonic problem of function $J_{O1}(t)$ and $J_{O2}(t)$ can be solved, but the large difference of distribution uniformity between object and background leads to relatively large threshold selection bias.

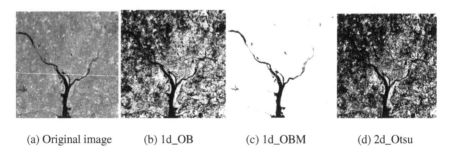

(a) Original image   (b) 1d_OB   (c) 1d_OBM   (d) 2d_Otsu

Fig. 5. The thresholded results of aerial image

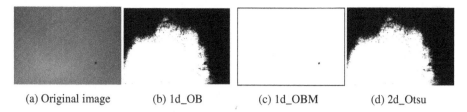

| (a) Original image | (b) 1d_OB | (c) 1d_OBM | (d) 2d_Otsu |

Fig. 6. The thresholded results of cont image

Therefore, for such images, the threshold formula need to be adjusted combining with the other spatial information.

Based on gray distribution histogram of image, the optimal threshold should be located at the valley point, therefore, using the idea of revised valley-emphasis [13], taking into account the neighborhood information of the valley point, a modified thresholding method based on the relative homogeneity is proposed.

Assuming the original image histogram is denoted as $f(g)$, then the modified formula is:

$$\bar{f}(g) = [f(g-m) + f(g-m+1) + \ldots + f(g) + f(g+1) + \ldots + f(g+m)] \quad (11)$$

This formula can be viewed as a filter, it can be simplified as:

$$\bar{f}(g) = \sum_{i=-m}^{m} f(g+i) \quad (12)$$

In which, m can be valued 1,2,3. The greater m, the longer the filter.

Since the optimal threshold should be located at the valley point, the gray probability and the neighborhood gray probability should be taken as the minimum. So a new threshold formula is constructed as:

$$J_{OBM}(t) = \bar{f}(t) * [J_{OB}(t)] \quad (13)$$

Table 1. Comparison of segmentation results for 4 methods on 2 images

Method	1d_O1	1d_O2	1d_OB	1d_Otsu
Rice	42	227	117	124
Letter	2	253	155	150

Table 2. The threshold results of 3 methods

Method	1d_OB	1d_OBM	2d_Otsu
Aerial	135	48	145
Cont	149	75	148

The optimal thresholding value t^* is:

$$t^* = Arg \min_{0 \leq t \leq L-1} [J_{OBM}(t)] \tag{14}$$

The results of valley point histogram modification method is shown in Figs. 5 and 6 (c), it is simplified as 1d_OBM. Figures 5 and 6(a) are original images, Figs. 5 and 6(b) and Figs. 5 and 6(d) are the results by 1d_OB and 2d_Otsu methods. From Table 2, we can see the thresholded values. For the two images, there is large difference between two classes. The results show that the integrity and clarity of object extraction are obviously improved by using the modified method. In this paper, we take parameter $m = 2$.

6 Conclusion

In this paper, we discussed thresholding method based on uniform classification criterion, analyzed the limitations of Otsu's method. Based on the method focusing on the objects, using gray and neighborhood average gray histogram, a more detailed description of the threshold discriminant function is developed. Taking into account the homogeneity information both of foreground and background, the threshold discriminant criterion based on the relative homogeneity is proposed.

Comparing with the method focusing on the objects, this method introduced the relative uniformity information, the adaptability is better to images. For some images with specific histogram distribution, using the idea of neighborhood histogram modification, the optimal threshold point should be located at the valley point. The method makes a modification to the deviation because of the different uniformity, and the segmented result is more reasonable.

Evaluation of the application to several test images illustrates the adaptability of the proposed methods, and which is better than 1d_Otsu, Chen's method and 2d_Otsu. It is obviously forward to extend the methods to multilevel thresholding problem by terminating the grouping process as the expected segment number is achieved.

References

1. Bardera, A., Boada, I., Feixas, M., Sbert, M.: Image segmentation using excess entropy. J. Signal Process. Syst. **54**(1–3), 205–214 (2009)
2. Frery, A.C., Jacobo-Berlles, J.J., Gambini, J., Mejail, M.E.: Polarimetric SAR image segmentation with B-splines and a new statistical model. Multidimension. Syst. Signal Process. **21**(4), 319–342 (2010)
3. Pal, N., Pal, S.: A review on image segmentation techniques. Pattern Recogn. **26**(9), 1277–1294 (1993)
4. Sezgin, M., Tasaltin, R.: A new dichotomization technique to multilevel thresholding devoted to inspection applications. Pattern Recogn. Lett. **21**(2), 151–161 (2000)
5. Sezgin, M., Sankur, B.: Survey over image thresholding techniques and quantitative performance evaluation. J. Electron. Imaging **13**(1), 146–165 (2004)

6. Otsu, N.: A threshold selection method from gray-level histograms. IEEE Trans. Syst. Man Cybernet. **9**(1), 62–66 (1979)
7. Hou, Z., Hu, Q., Nowinski, W.: On minimum variance thresholding. Pattern Recogn. Lett. **27**(14), 1732–1743 (2006)
8. Songcan, C., Daohong, L.: Image binarization focusing on objects. Neurocomputing **69**(16–18), 2411–2415 (2006)
9. Kwon, S.H.: Threshold selection based on cluster analysis. Patten Recogn. Lett. **25**(9), 1045–1050 (2004)
10. Fanyan, N., Yonglin, W., Pan, M.: Two-dimensional extension of variance-based thresholding for image segmentation. Multidimension. Syst. Signal Process. **24**(3), 485–501 (2013)
11. Fanyan, N., Jianqi, L., Tianyi, T.: Image segmentation using two-dimensional extension of minimum within-class variance criterion. Int. J. Signal Process. Image Process. Pattern Recogn. **6**(5), 13–24 (2013)
12. Girdhar, A., Gupta, S., Bhullar, J.: Weighted variance based scale adaptive threshold for despeckling of medical ultrasound images using curvelets. J. Med. Imaging Health Inform. **5**(2), 272–281 (2015)
13. Jiulun, F., Bo, L.: A modified valley-emphasis method for automatic thresholding. Pattern Recogn. Lett. **33**(6), 703–708 (2012)

Innovative Technology
and Applications

Searching of Circular Neighborhoods in the Square and Hexagonal Regular Grids

Vojtěch Uher$^{(\boxtimes)}$, Petr Gajdoš, and Václav Snášel

Department of Computer Science and National Supercomputing Center,
VŠB-Technical University of Ostrava, Ostrava, Czech Republic
{vojtech.uher,petr.gajdos,vaclav.snasel}@vsb.cz

Abstract. The nearest neighbors (NN) algorithm is a traditional algorithm utilized in many research areas like computer graphics, classification and/or machine learning. This paper aims at the fixed-radius nearest neighbors algorithm in 2D space. The neighbors are searched within a circular neighborhood which is positioned in the bounded space. The radius of the circle is known in advance. The uniform grids can be efficiently utilized for a nearest points query acceleration. This paper presents a study comparing the square and the hexagonal uniform grids and their suitability for a circular neighborhood querying. The two algorithms checking the mutual position/intersection of a circle and a square or a hexagon are described. The tests show the supremacy of the hexagonal grid.

Keywords: Nearest neighbors · NN · Hexagon · Uniform grid · Hexagonal grid

1 Introduction

The fixed-radius nearest neighbors (FRNN) algorithm searches all the points in a circular neighborhood with the defined radius. Usually, the neighbors of all vertices of a dataset are searched. The brute force algorithm thus computes $\mathcal{O}\left(|V|^2\right)$ distances, where $|V|$ is the total number of vertices. The radius is known in advance, so that it is possible to build a query structure that decides which spatial areas are worth to browse. Beside the traditional hierarchical clustering methods (Octree, kd-tree, BVH, etc.) [1–3], the straightforward methods utilizing the uniform regular grids can be used. The surrounding cells potentially hit by the radius contain the significant vertices. Franklin [4] pointed out that the hierarchical methods need a lot of memory to store the hierarchical information, their preprocessing time is mostly from $\mathcal{O}(|V|\log|V|)$ to $\mathcal{O}\left(|V|^2\right)$ and their query time complexity is about $\mathcal{O}(\log|V|)$. Thus, he introduced a nearest neighbor method called the Nearpt3 [4], which is based on the idea of the uniform grid and its query complexity is considered to be $\mathcal{O}(1)$. The Nearpt3 uses a linear representation of the vertices instead of the spatial trees. The number of vertices located in the surrounding cells is limited to some k. which usually represents just a small fraction of $|V|$. The computation time is thus much lower in the datasets with a reasonable vertex distribution. Several single-thread and parallel nearest neighbors algorithms based on the uniform grids have been published for large graphs visualization [5], particle

J. Pan et al. (eds.), *Intelligent Data Analysis and Applications*, Advances in Intelligent Systems and Computing 535, DOI 10.1007/978-3-319-48499-0_15

simulation [6, 7] and/or ray tracing [8]. The fundamental uniform grid cells are the squares, because they are very simple to use. However, the uniform hexagonal grids can be also utilized in a similar way. The paper [9] presents a method using a linear representation of the hexagonal cells for a search of the NN in 2D sparse datasets. Only the surrounding hexagonal rings are checked for neighbors. The hexagons can efficiently tile the 2D space and their shape is closer to a circle, so that they can even overcome the orthogonal grids in some cases. The papers [10–12] showed the advantages of the hexagonal grids for the circular neighborhood search on a theoretical level.

This paper aims at the FRNN search using the uniform grids. It provides a comparison of the orthogonal and hexagonal grids. The method introduced in this paper is based on the simple algorithm from the paper [9] which is improved here with novel algorithms testing the circle-square or circle-hexagon mutual position, which reduce the necessary number of distance computations.

2 Our Approach

This section introduces the modified FRNN algorithm. It extends the principle from [9] by the intersection algorithms. The two uniform grids are tested: square and hexagonal. First, the algorithms testing the intersection of a cell and a circle are described in Sect. 2.1. Next, the algorithm searching the FRNN is briefly introduced in Sect. 2.2.

2.1 Circle Intersection

This section presents two intuitive algorithms testing the mutual position of a circle and a cell (square, hexagon), which return the corresponding values. If a cell is completely outside the circle the cell will be skipped (Value 0). If a cell is just hit by the circle all the contained vertices will be sequentially checked (Value 1). If a cell is completely inside the circle all the contained vertices will be classified as neighbors (Value 2). The principal metric of the grid is the side size S of a square or a regular hexagon respectively. The advantage is that the both cell shapes can be precisely inscribed into a circle.

Circle-square Intersection. In the square grids, the intersection of the circle of neighborhood and the square cells have to be decided. The robust algorithm solving this problem was published in [13]. This section describes an algorithm specially simplified for the 2D space and its easy cases. The algorithm is illustrated in Fig. 1 and described in Function CircleSqrInt. First, the case when the distance of just one coordinate is too long is checked. Second, the cases when the whole circumscribed circle (purple) is out of the radius or completely inside the radius are tested. Next, the easy cases when the square is partially hit by radius from side (green area) and the cases when the radius hits the corners (red areas) are checked. Otherwise, no intersection is detected.

```
Function CircleSqrInt (C, P, halfSize, halfDiag, radius)
  Require
    C: center of a square
    P: center of a circular neighborhood
    halfSize: half of the square size S/2
    halfDiag: half of the precomputed square diagonal D/2
    radius: radius of the circular neighborhood
  Begin
    {vector connecting C and P}
    CPx = abs(Px - Cx)
    CPy = abs(Py - Cy)
    {no intersection easy cases}
    If CPx > (halfSize + radius) Then Return 0
    EndIf
    If CPy > (halfSize + radius) Then Return 0
    EndIf
    {no intersection and full intersection cases}
    dist = sqrt( (CPx)^2 + (CPy)^2 )
    If (dist - halfDiag) > radius Then Return 0
    EndIf
    If (dist + halfDiag) <= radius Then Return 2
    EndIf
    {partial intersection easy cases}
    If CPx <= halfSize Then Return 1
    EndIf
    If CPy <= halfSize Then Return 1
    EndIf
    {distance to the corners}
    If (dist - halfDiag) <= radius Then Return 1
    EndIf
    Return 0
  End
```

Circle-hexagon Intersection. This section describes an algorithm checking the intersection of a circle and a hexagon. This is utilized for a neighborhood search in the 2D hexagonal grids. The pointy-topped hexagonal grid is considered. The algorithm is illustrated in Fig. 2 and described in Function CircleHexaInt. First, the intersection with the gray bounding box is checked. Next, if the whole circumscribed circle (purple) is inside the radius, the full intersection is detected. If it is not even hit by the radius, no intersection is detected. If the green inscribed circle is hit by the radius, a partial intersection is detected. Otherwise, the area between the both circles have to be checked according to distance d computed from the angle between the line connecting the C and P and the grid orientation $u = (0, 1)$. The modified algorithm could be used for a general orientation of the grid. Equations would be changed according to a different bounding box and orientation *u*.

```
Function CircleHexaInt (C, P, halfWidth, radius)
  Require
      C: center of a hexagon
      P: center of a circular neighborhood
      u = (0, 1): the direction of the hexagonal grid
      halfWidth: half of the hexagon width w/2
      radius: radius of the circular neighborhood
  Begin
      {vector connecting C and P}
      CPx = abs(Px - Cx)
      CPy = abs(Py - Cy)
      {no intersection easy cases}
      If CPx > (halfWidth + radius) Then Return 0
      EndIf
      If CPy > (size + radius) Then Return 0
      EndIf
      {no intersection and full intersection cases}
      dist = sqrt( (CPx)^2 + (CPy)^2 )
      If (size + radius) < dist Then Return 0
      EndIf
      If (radius - size) >= dist Then Return 2
      EndIf
      {partial intersection easy case}
      If (halfWidth + radius) >= dist Then Return 1
      EndIf
      {area between the circles}
      alpha = acos(CPy/dist)
      If alpha > pi/3 Then
        If alpha > 2pi/3 Then alpha = alpha - 2pi/3
        Else alpha = alpha - pi/3
        EndIf
      EndIf
      beta = 2pi/3 - alpha
      d = size * sqrt(3/2) * 1/sin(beta)
      If (d + radius) >= dist Then Return 1
      EndIf
      Return 0

  End
```

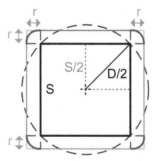

Fig. 1. Computation of the circle-square intersection. The green square represents the area where the radius can hit the square cell. The purple circle is the area which is tested for the full intersection with the radius. The red corners represent the areas where the point to corner vertex distance has to be checked. (Color figure online)

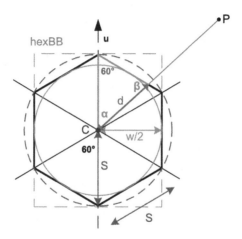

Fig. 2. Computation of the circle-hexagon intersection. The gray rectangle represents the area where the radius can hit the hexagonal cell. The purple circle represents the area which is tested for the full intersection with the radius. If the radius hits the inscribed green circle, a partial intersection is detected. The red distance d is computed to handle the area between the two circles (Color figure online).

2.2   Fixed-Radius Nearest Neighbors Search

The FRNN algorithm using the presented intersection algorithms is briefly described here. A point cloud is usually a set of unordered points. This set has to be divided to clusters, which represent a generalized location of contained vertices. The clusters are the grid cells here. The FRNN algorithm is based on the searching procedure introduced in [9] which is accelerated by the intersection algorithms. The linear vertex representation based on a list of existing cells (clusters) is utilized to address the surrounding cells on the grid hit by the radius r (details in [9]).

Initialization. The $|V|$ vertices have to be localized in the grid. For each vertex coordinates of the cell containing the vertex have to be computed. The vertex localization is trivial in the orthogonal grid. The localization in the hexagonal grid is explained e.g. in [9]. The cell coordinates are recalculated to a linear hash according to the C-curve order. The vertices are sorted then according to these hashes. The point clouds are usually sparse, thus this kind of clustering secures that only the existing cells are considered. A list Cs of cells is built by grouping together of the vertices of the same cell.

Fixed-radius Nearest Neighbors Query. A query vertex p is localized in the grid. The algorithm spirals out to the surrounding rings of cells containing the potential neighbors (see the indexations in Fig. 3). The precomputed list Cs of clusters is used to address the cells. The list of neighbors Rs is returned. The searching algorithm goes as follows:

1. A query vertex **p** is localized in the grid. The coordinates and the linear hash of the corresponding cell are computed.
2. The surrounding cells are browsed in the manner described in Fig. 3.
3. Each cell is checked for circle-cell intersection. The algorithm distinguishes the three cases depending on the value returned by an intersection function: 0 – no intersection, 1 – partial intersection, 2 – full intersection (see Sect. 2.1)
4. In the cases of Value 1 and 2, the cell is binary searched in the Ls according to the linear hash and the contained vertices are proceeded and popped to the Rs.
5. The list of all neighbors Rs within the fixed radius r is returned.

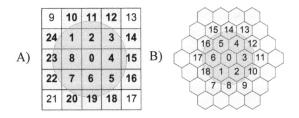

Fig. 3. Indexation of the cells: (A) Square, (B) Hexagonal

The presented FRNN algorithm browses the c surrounding cells and checks maximally n vertices in each cell. The number of cells in the list Cs is $|Cs| \leq |V|$. In the worst case, the query time is $\mathcal{O}(c \cdot \log_2 |Cs|) + \mathcal{O}(c \cdot n)$ if all the surrounding cells are hit by the radius. However, the real time complexity will be much lesser. See the Sect. 3.

3   Experiments

This section shows the comparison of the square and hexagonal grids. Both grids were tested on the artificial datasets with $|V| = 20000$ shown in Fig. 4. The charts in Fig. 5 show the comparison of the grids and their suitability for the circular neighborhood

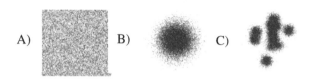

Fig. 4. Random vertex distributions: (A) Uniform, (B) Gaussian, (C) Gaussian islands

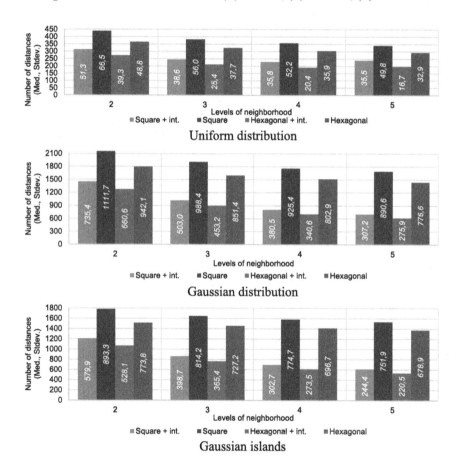

Fig. 5. Medians of distances calculated to find the FRNN of $|V|$ vertices on the square and hexagonal grids. The neighbors were searched with/without application of the circle-cell intersection function and different grid segmentation. The white numbers represent the standard deviation.

search. The Euclidean distances calculated to find the FRNN of all the $|V|$ vertices are compared. The neighborhood is divided to the specific number of levels (cell rings) according to the grid segmentation. The FRNN methods with and without application of the presented intersection algorithms are compared. The dependency on the grid segmentation is shown as well. The approaches utilizing the intersection algorithms reflect the position of a vertex in the cell and they overcome the previous basic algorithm [9].

4   Conclusion

The Fixed-Radius Nearest Neighbors (FRNN) algorithms comparison on the square and hexagonal grids was presented. The two simple algorithms to check the circle-square and circle-hexagon intersection were described. The experiments showed that the intersection algorithms significantly reduced the number of computed distances in comparison with the basic approach presented in [9]. The hexagonal grid is obviously better designed for the circular neighborhoods search than the square one. The gentler segmentation of the grids also reduced the number of distances on the tested datasets.

Acknowledgment. This work was supported by The Ministry of Education, Youth and Sports from the National Programme of Sustainability (NPU II) project "IT4Innovations excellence in science - LQ1602". This work was supported by SGS project, VSB-Technical University of Ostrava, under the grants no. SP2016/97 and no. SP2016/68.

References

1. Samet, H.: The quadtree and related hierarchical data structures. ACM Comput. Surv. **16**(2), 187–260 (1984)
2. Bédorf, J., Gaburov, E., Zwart, S.P.: A sparse octree gravitational N-body code that runs entirely on the GPU processor. J. Comput. Phys. **231**(7), 2825–2839 (2012)
3. Yin, M., Li, S.: Fast BVH construction and refit for ray tracing of dynamic scenes. Multimedia Tools Appl. **72**(2), 1823–1839 (2014)
4. Franklin, W.R.: Nearest Point Query on 184,088,599 Points in E^3 with a Uniform Grid (2006). http://wrfranklin.org/Research/nearpt3/
5. Uher, V., Gajdoš, P., Ježowicz, T.: Solving nearest neighbors problem on GPU to speed up the Fruchterman-Reingold graph layout algorithm. In: 2015 IEEE 2nd International Conference on Cybernetics (CYBCONF), pp. 305–310. IEEE Press (2015)
6. Green, S.: Particle simulation using cuda. NVIDIA Whitepaper **6**, 121–128 (2010)
7. Hoetzlein, R.: Fast fixed-radius nearest neighbors: interactive million-particle fluids. In: GPU Technology Conference (GTC) 2014 (2014)
8. Kalojanov, J., Slusallek, P.: A parallel algorithm for construction of uniform grids. In: 2009 Proceedings of the Conference on High Performance Graphics, HPG 2009, pp. 23–28. ACM, New York (2009)

9. Uher, V., Gajdoš, P., Ježowicz, T., Snášel, V.: Application of hexagonal coordinate systems for searching the K-NN in 2D space. In: Snášel, V., Abraham, A., Krömer, P., Pant, M., Muda, A.K. (eds.) Innovations in Bio-Inspired Computing and Applications. AISC, vol. 424, pp. 209–220. Springer, Heidelberg (2016)
10. Lester, L.N., Sandor, J.: Computer graphics on a hexagonal grid. Comput. Graph. **8**(4), 401–409 (1984)
11. Haverkort, H.J.: Recursive tilings and space-filling curves with little fragmentation. CoRR abs/1002.1843 (2010)
12. Ben, J., Tong, X., Chen, R.: A spatial indexing method for the hexagon discrete global grid system. In: 2010 18th International Conference on Geoinformatics, pp. 1–5, June 2010
13. Ratschek, H., Rokne, J.: Test for intersection between circle and rectangle. Appl. Math. Lett. **6**(4), 21–23 (1993)

A Moving Object Detection Algorithm Based on a Combination Optical Flow and Edge Detection

Yao Xi, Ke-bin Jia$^{(\boxtimes)}$, and Zhong-hua Sun

Multimedia Information Processing Group,
College of Electronic Information and Control Engineering,
Beijing University of Technology, Beijing, China
kebinj@bjut.edu.cn

Abstract. Moving objectives detection is the basis for the application of computer vision technology, in intelligent video surveillance systems in many areas have a wide range of applications for dynamic in the context of the moving object detection, presented in this paper that will improve the pyramid optical flow law and the combination of edge detection moving object detection method. First of video pre-processing applications, Wavelet Threshold to remove noise; second, through the pyramid optical flow to achieve maximum displacement appears moving objectives; then uses a canny edge detection for the edge of an image information; and finally will get the optical flow of information and the edge of the characteristics of the information so the second value after the integration of the morphology through the follow-up to the campaign to get accurate target information. Lab results indicate that the algorithm real-time high, the versatile good, you can more accurately detect moving objectives.

Keywords: Wavelet threshold de-noising · Moving background · Pyramid Lucas-Kanade optical flow · Canny edge detection

1 Introduction

At the continuous development of economic growth in China under the background of China's industrial transformation is becoming increasingly urgent that the German Government in 2013 put forward the strategy of industrial 4.0. Industrial 4.0 is intelligent manufacturing-oriented revolutionary production methods, aimed at the adoption of communications technologies, Virtual networks, and the physical network entities combine intelligent manufacturing industry to economies in transition. This transition has been the Chinese Government attach great importance to the 2014 12, "Made in China 2025" concept, which was first introduced. Industrial Shop Digital is to achieve the transformation of industries and essential step. So far, China has many digital people-less factories. At the same time, the digital industrial shop security monitoring is also a very important issue, modern people for video surveillance, no longer stay in the manual operation of the auxiliary visualization level, but rather through the computer

© Springer International Publishing AG 2017
J. Pan et al. (eds.), *Intelligent Data Analysis and Applications*, Advances in Intelligent Systems and Computing 535, DOI 10.1007/978-3-319-48499-0_16

automated controls to identify the tracking of the movement in the video, the intelligent monitoring.

In accordance with the scene with the camera or the existence of relative movement video processing can be split into static background of sports under Target Detection and dynamic movement in the context of the target detection. Static background of sports under target detection now have more sophisticated algorithms: *Frame Difference Methods; background subtraction;* optical flow [1]. In relation to the static background of dynamic background has substantially increased movement target detection of complexity. Therefore, in the context of the dynamic movement is a key target detection is also a difficult point.

For dynamic background using Campaign target detection methods template matching [2], background motion compensation [3] and the optical flow. Template matching of thinking is to obtain the aims of the a priori characteristics, and make use of their training classifier, through these classifier vote to detect target. In the case of known campaign objectives a priori characteristics, you can use this method. Based on the idea of the motion compensation will be in the video from the camera of the Global Campaign for movement and by the movement of local sports. The global campaigns and local sports are two relative independence movement so that it is accessible through the camera from the motion compensation to eliminate the global movement to detect the movement of local campaign objectives. Optical Flow [4] may not know any a priori features detect the movement of an object is an application prospects are better algorithm, but this method is not able to detect the movement of the complete contour and vulnerable to light and the impact of noise.

This article examines the camera movement that improved optical flow approach with edge detection of the aims of the methods of detecting, *Horn–Schunck* method [5] for calculating the simple geometric significance intuitive, but when the speed is very high, it is based on the assumption that the grayscale remain there is a large error; L-k optical flow requirements in a smaller space and maintains a constant motion vectors, and pyramid optical flow [6] through sampling and the establishment of multi-tiered pyramid structures layer can be maximum displacement between the images in the adjacent movement, displacement has become small enough that this article use L-k pyramid optical flow approach with canny edge detection method [7], quick operation speed for maximum displacement robot motion and more accurate detection of dynamic context of moving objects.

2 Image Pre-processing

Digital video in the acquisition, encoding, transfer, in the process of decoding will inevitably introduce various noise. Image cleanup of commonly divided into empty domain noise reduction and frequency domain cleanup of two parts. But these methods only in an empty field or frequency domain has local capacity to analyze in inhibiting the image noise, would compromise the edge of an image in the image details in the following noise reduction become blurred.

This article uses wavelet threshold [8] of the methodologies for the clean-up of image noise reduction (Figs. 1 and 2).

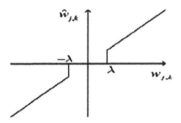

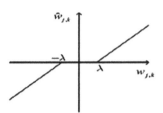

Fig. 1. Hard-thresholding   **Fig. 2.** Soft-thresholding

2.1   Wavelet Threshold de-Noising Method

(1) hard-thresholding

$$\hat{\omega}_{j,k} = \begin{cases} \omega_j, k & |\omega_j, k| \geq \lambda \\ 0 & |\omega_j, k| < \lambda \end{cases} \tag{1}$$

(2) soft-thresholding

$$\hat{\omega}_{j,k} = \begin{cases} sgn(\omega_j, k) \cdot (|\omega_j, k| - \lambda) & |\omega_j, k| \geq \lambda \\ 0 & |\omega_j, k| < \lambda \end{cases} \tag{2}$$

These two functions have been widely used in practice and achieved good results, but they also have same shortcomings. In hard threshold function ω_j, k is discontinuous in the λ *and* $-\lambda$ so reconstruction image will appear visual distortion, but the results use soft threshold method will be futher smoother. so this paper adopt the soft threshold method.

3   The Research Methods of This Paper

3.1   The Basic Principle of Optical Flow

Optical flow field can be simply understood as the velocity vector field of the object, according to the gray invariance assumption, we can get the basic process of optical flow.

Let us assume that the gray values of point (x,y) is I(x,y,t) at time t, the horizontal and vertical component of optical flow w(u, v) are u(x,y)和v(x,y):

$$u = \frac{dx}{dt} \qquad v = \frac{dy}{dt} \tag{3}$$

Interval for dt, point I(x + dx, y + dy, t + dt), dt → 0, maintaining the same gray, The following result can be obtained I(x,y,t) = I(x + dx, y + dy, t + dt) and finally get optical flow constraint equation:

$$-\frac{\partial I}{\partial t} = \frac{\partial I}{\partial x}u + \frac{\partial I}{\partial y}v \tag{4}$$

The relationship between the gradient of the image and the optical flow velocity is represented by the image. Because the optical flow constraint equation is only one equation, there are two unknowns, so we need to increase other constraints to solve this problem.

3.2 Improved L-K Pyramid Optical Flow Algorithms

L-k optical flow Act originally submitted in 1981, the assumption that the algorithm in a smaller space within the neighborhood motion vectors constant use of weighted least squares to estimate the optical flow. As a result of the algorithm is applied to the input image on a set of points is easier if it is widely used in sparse optical flow farm. L-k optical flow law based on the following three assumptions:

(1) The brightness is constant. For a grayscale image, on the assumption that the entire tracked period, pixel brightness unchanged.
(2) The time continuous or movement is a small movement. Image of the Movement with respect to time can be relatively slow, real applications, point of time relative to the image in the moving to a small enough so that the objectives in the adjacent frames between the movement on the relatively small.
(3) Space. The same scene. The same adjacent points on the surface of the movements of the similarity to these points in the image's projection on also in neighboring regions.

Optical flow error is defined as:

$$\sum_{(x,y)\in\Omega} W^2(x)(I_x u + I_y v + I_t)^2 \tag{5}$$

$W^2(x)$ present the weight function of the window, The function makes the influence of the neighborhood center region on the constraint, which is larger than that of the peripheral region.

Get the solution:

$$U = (A^T W^2 A)^{-1} A^T W^2 B \tag{6}$$

t time:

$X_i \in \Omega$, $A = [\nabla I(X_1), ..., \nabla I(X_n)]^T$, $W = \text{diag}[W(X_1), ..., W(X_n)]$, $B = -[[I(X_1), ..., I(X_n)]^T]$.

L-k algorithm is based on slow to change the brightness of the image also assumes that the so the algorithm applies only to a small image shift movement in maximum displacement of time we cannot achieve accurate optical flow estimates. For this problem, Pyramid optical flow law had been submitted through the creation of the pyramids, combining L-k optical flow method in multi-scale to calculate the optical flow, you can increase the accuracy of the Optical Flow calculation.

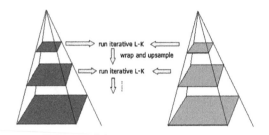

Fig. 3. Pyramid structure

The pyramids optical flow of ideas is on each frame within the image down is hierarchical and the establishment of multi-tiered pyramid structure, as long as the image to a hierarchical small enough maximum displacement between the images in the adjacent Campaign will also be small enough, this time from the top layer started using L-k optical flow method calculates the aims of the optical flow will be calculated from the optical flow to the underlying projections, then calculate the next lower level of optical flow, until the final estimate of the original image optical flow (Fig. 3).

The calculation steps are as follows:

(1) Use the original image as the base of the pyramid (L = 0), sampling at different resolution levels according to accuracy, When the image is divided into a certain level, the larger displacement parameters can be small enough to satisfy the constraints of optical flow computation.

(2) Calculation from the top-level to bottom. suppose the Initial optical flow vector of L+1 is g_L, Δf_L is the result of optical flow at L+1 layer, Mapping relation between levels:

$$g_{L-1} = k(g_L + \Delta f_L) \qquad (7)$$

(3) The total number of layers in Pyramid is N, optical flow of top-level is $g_{N-1} = 0$ Motion parameters of original image as follow:

$$f = g_0 + \Delta f_0 = \sum_{L=0}^{N-1} K_L \Delta f_L \qquad (8)$$

The results are as follows (Figs. 4 and 5):

What we study is the moving object detection under the dynamic background, the position of the target is constantly changing in the image, and the background and the target position are constantly changing in the video image. After the formation of the optical flow field of each frame image in video sequence, the moving object can be extracted from the dynamic background by clustering algorithm [9].

Fig. 4. Result of corner detection   **Fig. 5.** Optical flow vector of corner

3.3 Canny Edge Detection Algorithm

In a large number of edge detection algorithm in the 1986 canny-based optimization algorithms canny edge detection is to a very good signal-to-noise ratio and accuracy, which has received wide use, mainly because it offered by far the most stringent definition for edge detection for the three standard, in addition to its relatively simple algorithm causes the entire calculation process can be in a relatively short period of time.

Canny research the best edge detection, which, given the characteristics of the evaluation of the performance of the strengths and weaknesses of the Edge Detection three targets:

(1) Good signal-to-noise ratio, the forthcoming non-edge points to the edge points and edge points awarded for non-edge is the probability to low;

(2) Better positioning performance, or test the edge points to the extent possible, at the edge of the actual center;

(3) On the Single Edge only has a unique response, i.e. the individual edge produce multiple responses to low probability and false response border should be maximum suppression (Figs. 6 and 7).

Fig. 6. Original image   **Fig. 7.** Canny edge detection

The concrete implementation steps are as follows:

(1) A Gaussian filtering to Noise Reduction;
(2) Using a first-order differential is calculated in horizontal and vertical gradient amplitude component to be images of the gradient amplitude M and the direction of the gradient e, in order to increase the precision of the gradient of each pixel in the image points use sobel operator is calculated as local gradient amplitude values and the edges. The gradient direction amplitude values from the point of greatest for the edges of the image;

$$M = \sqrt{G_x{}^2 + G_y{}^2} \tag{9}$$

$$\theta = \arctan(G_y/G_x) \tag{10}$$

(3) Thinning image edge;
(4) Capture edge data

The results are as follows:

3.4   Experimental Result

After processing the Optical flow information and edge wo have got[10], and fuses binary information we have got, Finally get the information of the moving target.。 In this paper, the "and" operation is used to fuse the binary information.

$$F_{np}(i,j) = \begin{cases} 1 & \text{if } P_{np}(i,j) = 1 \text{ and } L_{np}(i,j) = 1 \\ 0 & \text{else} \end{cases} \tag{11}$$

$F_{np}(i,j)$ stands for the image after fusing the binary information, $P_{np}(i,j)$ stands for the binary information of the edge, $L_{np}(i,j)$ stands for the binary information of the optical flow

"And" operation can get rid of the extra information except for the moving target, make the contour of the moving target more clear, and keep both information;reinforcing motion information; improve the accuracy of detection.

And morphological processing [11] could process the images using Dilation and erosion.

Experimental results show that the proposed algorithm has good real-time performance and good general performance because of we using the Pyramid LK optical flow and Canny edge detection algorithm, the algorithm can accurately detect the moving target which can be seen from Fig. 8.

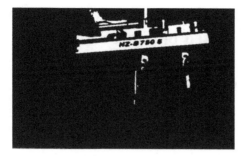

Fig. 8. Final result

4 Conclusion

Presented in this paper that improve the pyramid L-K optical flow and the combination of edge detection moving object detection method. Lab results indicate that the algorithm real-time high, the versatile good, you can more accurately detect moving objectives.

References

1. Zhang, M.-J., Yu, W.-J., Yuan, Z., Huang, Z.-J.: Research of video-based object detection algorithm. Comput. Eng. Softw. **04**, 40–45 (2016)
2. GAO, Y.-P., Ming-Yi, H.E.: Moving target detection combined two frame differences with template matching methods under dynamic background. Electron. Des. Eng. **05**, 142–145 (2012)
3. Sun, B., Huang, S.-Z.: Target detection and tracking under moving background. J. Electron. Measure. Instrum. **3**, 206–210 (2011)
4. Lucas, B.D., Kanade, T.: An iterative image registration technique with an application to stereo vision. In: IJCAI 1981 (1981)
5. Xie, J.-B., Wang, H., Cheng, J.-H., Liu, T.: Optical flow computation method based on HS constrain t and outline condition. Syst. Eng. Electron. **4**, 761–763 (2009)
6. Diamantas, S.C., Oikonomidis, A., Crowder, R.M.: Depth computation using optical flow and least squares. In: Proceedings of 2010 IEEE/SICE International Symposium, pp. 7–12. IEEE Prees, New York (2010)
7. Bao, P., Lei, Z., Xiao-lin, W.: Canny edge detection enhancement by scale multiplication. IEEE Trans. Pattern Anal. Mach. Intell. **27**(9), 1485–1490 (2005)
8. Chen, X.-X., Wang, Y.-J., Liu, L.: Deep study on wavelet threshold method for image noise removing. Laser Infrared **01**, 105–110 (2012)
9. Wang, Q., Wang, C., Feng, Z.-Y., Jin-Feng, Y.E.: Review of K-means clustering algorithm. Electron. Des. Eng. **07**, 21–24 (2012)
10. Liu, Y.-H., Wang, Z.-F., Yang, J.-Y., Xie, Z.-X.: A new color image binarization method and its application. Chin. J. Med. Phys. **01**, 3873–3876+3897 (2013)
11. Wu, Y.: Application of morphology in image processing. JI SUAN JI YU XIAN DAI HUA **05**, 90–94 (2013)

The Real-Time Depth Map Obtainment Based on Stereo Matching

Fei Wang, Kebin Jia, and Jinchao Feng[✉]

Multimedia Information Processing Group, College of Electronic Information
and Control Engineering, Beijing University of Technology, Beijing, China
fengjc@bjut.edu.cn

Abstract. The depth measurement according stereo vision is a very popular method of measuring the depth information. But the calculation in stereo matching is large, time-consuming and in large errors in matching, when used in real-time systems to obtain depth information are often ineffective. This paper improves one of the local matching algorithms, builds a complete real-time system for calculating depth image with different SAD (Sum of Absolute Differences) matching window in different texture regions. In this paper, we capture images from the Bumblebee2 stereoscopic camera which mounted on a small unmanned car, then use Matlab calibrating the camera and do the calculation to correct the raw images and do stereo matching in VS2010 to get the depth map. This method is simple, real-time performance and adaptability, and the quality of the depth map calculated from this method is somewhat improved.

Keywords: Depth information · Stereo matching · Texture · SAD · Real-time

1 Introduction

With the development of the robot industry, unmanned car industry and virtual reality industry, three-dimensional reconstruction of machine vision has become more and more popular. Depth information acquired is the most basic and most important part of three-dimensional graphics. The main approaches for depth acquisition include laser scanning, structured light and stereo [1]. Laser scanning, also called ToF (Time of Flight), rangefinders measure the time it takes for the light to travel to the objects and back. This method can get the precise, distant data, but the device is heavy, not flexible, large and expensive. The method of structured light uses a projector to illuminate the object with the structured light and get the back information. The optical encoder technology that used in Kinect is a kind of structured light. This approach can get accurate depth data but just for limited range and is vulnerable to the outside light. The stereo imaging method used in this paper calculates the depth of the image by two images at different angles. This approach is simple, flexible, affordable, of course, also can get the accurate data. The camera calibration is easier and more precise, after Zhang Zhengyou proposed camera calibration method, so the main issues on stereo vision lies on stereo matching. In the efforts of many scholars, the performance of stereo matching is better and better.

© Springer International Publishing AG 2017
J. Pan et al. (eds.), *Intelligent Data Analysis and Applications*, Advances in Intelligent Systems and Computing 535, DOI 10.1007/978-3-319-48499-0_17

In the 1980s, a visual computing theory proposed by Marr applied to the binocular stereo matching started the exploration of stereo vision theory. Today, stereo matching can be divided into global and local stereo matching. In local stereo matching, Kim et al. [2] described the applications of variable window algorithm, as set forth correlation function improves the matching precision on depth discontinuity area. Yoon et al. [3] proposed an adaptive weighted cost of window aggregation method, which firstly to calculate a weight for each pixel in the window. The weight for pixel is depended on the chromatic aberration and spatial position difference between the current pixel and center pixel. This method can get high-quality disparity map. Because of the large shape of the selection window, and the high complex weight calculation, the performance of the algorithm is not so good. The algorithm that Zhang et al. [4] proposed distributes for each pixel horizontal and vertical two arms that orthogonal itself. This algorithm also can get high quality disparity map, but need compare the color of center pixel with any other pixel which cost a lot of time and can't satisfy the real-time requirement. The global matching algorithms usually used are dynamic programming stereo matching method, graph cuts stereo matching method and the belief propagation matching method. In this paper, we improve the BM (Block Matching) algorithm in OpenCV which belongs to local stereo matching. BM algorithm uses fixed SAD window for stereo matching, has a good real-time performance. The proposed algorithm in this paper firstly extracts the edge of images by Canny operator and then decides the size of SAD window according to the area (edge area or not) the pixel belongs to. The algorithm has low time complexity and good robustness, it improves the matching accuracy.

2 Camera Model and Calibration

The real world is a three-dimensional world. Despite some controversy in binocular stereo vision and monocular stereo vision, it is complicated and difficult to get the corresponding depth information relying on only one image. However, we can calculate the depth map easily through two images obtained from a calibrated stereo camera. In order to analyze images with geometry theory, we need to model the system of imaging, and then process them with geometry methods.

Four coordinate systems [5] are used in the stereo calibration include the world coordinate and camera coordinate, the image plane coordinate and pixel coordinate. Information in the world coordinate system switch to the camera coordinate system through an external calibration parameter matrix W (including rotation matrix R and translation matrix T), and then to image plane coordinate system through the inner calibration parameter matrix K. Assuming that P (X, Y, Z, 1) is a point in the world, and p (x, y, 1) is the corresponding point in the pixel coordinate, so we can get the equation $p = sKWP$ (s is the scale factor).

The point of the camera coordinate Pc (x$_c$, y$_c$, z$_c$), can be expressed as Pc = WP,

$$\begin{bmatrix} xc \\ yc \\ zc \end{bmatrix} = \begin{bmatrix} R & T \end{bmatrix} \begin{bmatrix} X \\ Y \\ Z \end{bmatrix} \tag{1}$$

where R represents a rotation matrix, T represents a translation matrix. This transformation just between two three-dimensional coordinates.

Inner calibration parameter matrix consist of camera focal length f and the center coordinate of imaging plane c. p(u, v) is a point in image plane, we can get the equation:

$$
\begin{bmatrix} u \\ v \\ w \end{bmatrix} = \begin{bmatrix} fx & 0 & cx \\ 0 & fy & cy \\ 0 & 0 & 1 \end{bmatrix} \begin{bmatrix} xc \\ yc \\ zc \end{bmatrix}
\tag{2}
$$

In this way, we can find the correspondence between image plane coordinate and world coordinate according to the two matrices. One of the most important purposes of calibration is to calculate the matrices and the other is to obtain the distortion coefficients of cameras.

Since calibration with Matlab [6] is simpler than OpenCV, and gains wider recognition, we use Zhang calibration method in Matlab to do stereo calibration. Firstly, use stereo camera to capture images of calibration target from different angles. Then input the images into Matlab for calibration and copy the data into VS2010 for more operation.

3 Image Correction

Image distortion [7] from camera is divided into radial distortion and tangential distortion, the former refers to the deviation distance of ideal pixel position and the actual pixel position to the center of the image, mainly caused by the lens surface defects; the latter refers to the deviation angle from the ideal pixel position and the actual pixel position in the polar coordinate system, mainly due to the lens and the imaging plane is not parallel. Wherein the radial distortion can be divided into negative radial distortion (barrel distortion) and a positive radial distortion (pincushion distortion).

Model the radial, tangential distortion and establish an objective function to fit it. The distortion model:

$$
\begin{aligned}
u' &= u(1 + K_1 r^2 + K_2 r^4 + K_3 r^6) + 2P_1 uv + P_2(r^2 + 2u^2) \\
v' &= v(1 + K_1 r^2 + K_2 r^4 + K_3 r^6) + 2P_2 uv + P_1(r^2 + 2v^2)
\end{aligned}
\tag{3}
$$

Where K represents radial distortion coefficient, P for tangential distortion coefficient, r denotes the radius which is $\sqrt{u^2 + v^2}$, and (u', v') is the ideal coordinate in the image plane coordinate.

Objective function:

$$
\min F = \sum_{i=1}^{N} (u - u')^2 + \sum_{i=1}^{N} (v - v')^2
\tag{4}
$$

This function uses least square method which is used to solve the nonlinear distortion model directly, and simplifies the problem of solving this. We can do this in camera calibration and get parameter series {k1, k2, k3, p1, p2}, then put them into VS2010.

We use the cvStereooRectify function in OpenCV to correct the camera inner parament and make imaging plane of camera in geometry is parallel with Bouguet epipolar constraint method. Then send the obtained parameters to cvInitUndistortRectifyMap (), and get undistort rectify map which can save time in gaining corrected images next time. Finally give the map to cvRemap () and redraw the corrected image.

4 Obtaining the Depth Map

Binocular depth calculation is based on the principle of parallax [8]. For a point in space, its position in the image will be different, due to different shooting angle from stereo camera. The distance Z could be calculated with triangle similarity principle if the two image planes of the stereo camera are parallel. As shown in Fig. 1, we can get the equation

$$\frac{||T|| - (x_l - x_r)}{Z - f} = \frac{||T||}{Z} \Rightarrow Z = \frac{||T|| f}{x_l - x_r} \tag{5}$$

where x_l, x_r are the abscissas of p_l, p_r, $||T||$ is the optical center distance which can be gained in the stereo calibration (Fig. 2).

Since the real-time requirements, local stereo matching method is used in the application. StereoBM algorithm calculates similarity with SAD, the biggest similarity point as the stereo matching point:

$$SAD(x, y, d) = \sum_{i=-m}^{m} \sum_{j=-n}^{n} |I_l(x+i, y+j) - I_r(x+i+d, y+j)| \tag{6}$$

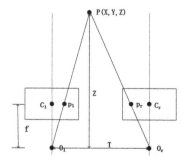

Fig. 1. Binocular camera imaging schematic

Fig. 2. Images before (left) and after (right) correcting

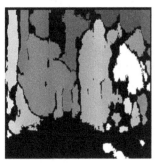

Fig. 3. Depth map from small (left), medium (middle), large (right) SAD window

where d is the parallax value. And d which can make the SAD as the minimum is the real parallax value (Fig. 3).

We can set the state value include pre-filter setting, SAD window size etc., and use findStereoCorrespondenceBM () to get the parallax value for calculating the depth.

By analyzing the depth maps above, the smaller SAD window can make the depth map have more edge information, but with more noise and mismatch in smooth area. With the increase of the SAD window size the effect on smooth area gets better, but the time program consumed gradually increased and the edge region gradually blurred. So this paper combines the idea of [9], using the Canny operator to get the edge of the image, and then processing the image with small SAD window in edge area and big SAD window in non-edge area (Fig. 4).

First of all, we use canny function to get the edge of single view, and then obtain the edge area map with mask. Finally, according to the map determine the size of sad window the current point to use, and get the final disparity and depth map. This test is done in different texture scenes, and compared with the depth image.

As shown in Fig. 5, the depth map in figure (a) is gained with the smaller SAD window and can be seen the depth information more accurate in edge portion where is rich of texture, but with more mismatching points in regions with lower degree of texture;

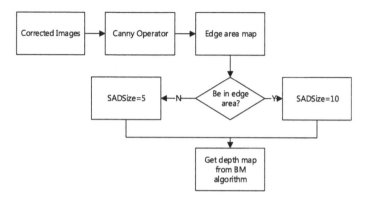

Fig. 4. Matching algorithm flowchart

    (a)   (b)   (c)   (d)

Fig. 5. General scene depth map

figure (b) is from larger SAD window, although the effect in low texture regions is better, many depth information for edge areas are lost, and cost a lot of time in matching; figure (c) is the optimal depth map from fixed SAD window; figure (d) is depth image obtained by the algorithm is proposed in this paper, and it has the best performance. We even can see chair behind the desk, and the floor is much better than others.

5   Conclusions

The system realizes the real-time acquisition of depth map with the binocular unmanned vehicle. To this end, the principle and implementation method of stereo camera calibration, image correction and stereo matching are researched. We calibrated the stereo camera in Matlab, and do image correction and acquisition of depth map in VS2010 with C++. This paper improves the original algorithm in OpenCV. The quality of depth map is improved on the basis of real-time and can adapt to different environment. But there is much space to improve the accuracy of the depth map, and need to continue exploring and researching.

Acknowledgments. This paper is supported by the Project for the National Key Technology R&D Program under Grant No.2011BAC12B03, the Key Project of Beijing Municipal Education Commission under Grant No.KZ201310005004, Basic Research Foundation of BJUT under Grant No.002000514314011 and Scientific Innovation Platform under Grant No.002000546615022.

References

1. Se, S., Jasiobedzki, P.: Stereo-vision based 3d modeling for unmanned ground vehicles. In: Proceedings of SPIE - The International Society for Optical Engineering, vol. 6561(1), 65610X-65610X-12 (2007)
2. Kim, G.B., Chung, S.C.: An accurate and robust stereo matching algorithm with variable windows for 3d measurements. J. Mechatron. **14**(6), 715–735 (2004)
3. Yoon, K.J., Kweon, I.S.: Adaptive support-weight approach for correspondence search. J IEEE Trans. Pattern Analy. Mach. Intell. **28**(4), 650–656 (2006)
4. Zhang, K., Lu, J., Lafruit, G.: Cross-based local stereo matching using orthogonal integral images. J IEEE Trans. Circ. Syst. Video Technol. **19**(7), 1073–1079 (2009)
5. Liu, F.C., Xie, M.H., Wang, W.: Stereo calibration method of binocular vision. J. Comput. Eng. Des. **32**(4), 1508–1512 (2011)
6. Wang, Z.Z., Zhao, L.Y., Liu, Z.Z.: Binocular stereo vision distance measurement system based on a combination of matlab and openCV. J. Tianjin Univ. Technol. **1**, 13 (2013)
7. Lin, H.Y., Jian, S., Liu, Y.M., Rong, C.: Camera calibration technique based on rectification of image aberration. J. Jilin Univ. **37**(2), 433–437 (2007)
8. Wang, H., Zhiwen, X.U., Xie, K., Jie, L.I., Song, C.: Binocular measuring system based on openCV. J. Jilin Univ. **2**, 13 (2014)
9. Guo, L.Y., Sun, C.Y., Zhang, G.Y., Wu, J.H.: Variable window stereo matching based on phase congruency. Appl. Mech. Mater. **380–384**, 3998–4001 (2013)

Handwritten Numbers and English Characters Recognition System

Wei Li[1], Xiaoxuan He[1], Chao Tang[2], Keshou Wu[1], Xuhui Chen[1],
Shaoyong Yu[1], Yuliang Lei[1], Yanan Fang[1], and Yuping Song[3(✉)]

[1] School of Computer and Information Engineering,
Xiamen University of Technology, Xiamen 361024, China
[2] Department of Computer Science and Technology,
Hefei University, Hefei 230601, China
[3] School of Mathematical Sciences, Xiamen University, Xiamen 361005, China
ypsong@xmu.edu.cn

Abstract. In this paper, we design a recognition system of the handwritten numerals and English characters based on BP neural network. In the system, we first make some preprocess to the image. Secondly, we extract the structural and statistical features of the image. Thirdly, we train a model on the data sets via BP neural network. Finally, we can predict the test image using the trained model. The experiments show that the handwritten numbers recognition rate can reach more than 94 % and the handwritten English characters is 44 % respectively. Besides that, the recognition rate of Handwritten number and English together is 78 %. It has been proved that our method is effective and robust.

Keywords: Handwritten character recognition · Image process · BP neural network

1 Introduction

With the rapid development of national science and technology, we come to the big data era. We need input the information of handwritten character in many places, such as ticket printing, automatic marking and so on (see [1–8]). As is well known these work will cost much time. So automation handwritten character processing is an attractive research direction. It will reduce the burden of those boring manual work [1, 6].

Character recognition technology is one of hotspots in the field of image processing and pattern recognition [5]. To satisfy the increasing demand of the society we have to process enormous data. It is particularly important to automatically use the computer on characters recognition. The research of this technology is very valuable in practice. Successful applications have been found in document digitization and retrieval, postal mail sorting, bankcheck processing, form processing, pen-based text input, and so on [7]. In this paper, we design a recognition system on digital and English character using the BP neural network.

J. Pan et al. (eds.), *Intelligent Data Analysis and Applications*, Advances in Intelligent Systems and Computing 535, DOI 10.1007/978-3-319-48499-0_18

2 Theory Algorithm

2.1 BP Neural Network

BP neural network [2] is based on the algorithm of error back propagation to train the multilayer feed forward network. This is one of the most popular method in the pattern recognition study. The topological structures of BP neural network model include input, hidden and output layer. As shown in Fig. 1, the BP network can learn and store a lot of mapping relation of input and output. It has two propagation process, respectively. One is the forward propagation and the other is reverse propagation.

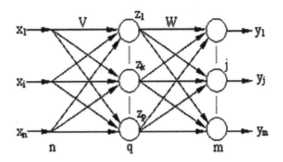

Fig. 1. Three layers topology structure of neural network.

2.2 Forward Propagation

Let n, m and q represent the number of nodes of the input layer, hidden layer and output layer respectively. The weight between the ith neuron of input layer and the jth neuron of hidden layer is V_{ki}. The weight between the jth neuron of hidden layer and the kth neuron of output layer is W_{jk}, as shown in Fig. 1.

The transfer function of hidden layer is $f_1(\cdot)$. The transfer function of the output layer is $f_2(\cdot)$ where $k = 1, 2, 3, \ldots, q, j = 1, 2, 3, \ldots, m$. And then the output of hidden layer nodes is

$$z_k = f_1\left(\sum_{i=0}^{n} V_{ki} x_i\right) \tag{1}$$

The output of output layer nodes is:

$$y_j = f_2\left(\sum_{k=0}^{q} W_{jk} z_k\right) \tag{2}$$

Thus, we have calculated the mapping relationship between the n-th and the m-th dimensional space vector.

2.3   Reverse Propagation

(1) Define Error Function

Let $X^1, X^2, \cdots, X^P$ be the inputting P samples. We denote y_j^p the outputting after the first p^{th} sample input to the BP neural network. As shown in Eq. (3), we can get the pth error

$$E_p = \frac{1}{2} \sum_{j=1}^{m} \left(t_j^p - y_j^p \right)^2 \tag{3}$$

where t_j^p is expected output.

For P samples, the global error is E. We can calculate E by the Eq. (4).

$$E = \frac{1}{2} \sum_{p=1}^{P} \sum_{j=1}^{m} \left(t_j^p - y_j^p \right) = \sum_{p=1}^{P} E_p \tag{4}$$

(2) Weight Variation of Output Layer

We make the global error E decrease by adjusting the weight W_{jk}. We use the accumulative algorithm to get it. That is as shown in the Eqs. (5)–(7).

$$\Delta w_{jk} = \frac{\partial}{\partial w_{jk}} \left(\sum_{p=1}^{P} E_p \right) = \sum_{p=1}^{P} \left(-\eta \frac{\partial E_p}{\partial w_{jk}} \right) \tag{5}$$

In the Eq. (5), η is the learning rate.

Finally the adjustment formula of the output layer weights of the each neuron is:

$$\Delta w_{jk} = \sum_{p=1}^{P} \sum_{j=1}^{m} \eta \left(t_j^p - y_j^p \right) f'_2 (S_j) z_k \tag{6}$$

(1) The weights of hidden layer

The adjustment of the hidden layer weights of each neuron formula is:

$$\Delta v_{ki} = \sum_{p=1}^{P} \sum_{j=1}^{m} \eta \left(t_j^p - y_j^p \right) f'_2 (S_j) w_{jk} f'_1 (S_k) x_i \tag{7}$$

2.4   Feature Extraction

In character recognition, there is a very important stage which is the feature extraction. Because the stand or fall of feature extraction can directly affect the final recognition accuracy. The key technology of digital image character recognition is to find effective character feature of algorithm, to select features with identifiable, independent, reliable, and features such as fewer properties.

In this paper we combine the structure features with the statistical features of character images. Among them, the structure feature of the structure extraction method is based on the character image. It can be the application of spatial structure to describe the feature of the strokes. And the statistical feature extraction method is to apply the lattice information of character image to do various transformations and to extract feature of the reentry. This method is based on the set according to the characters of image pixels.

3   System Design

The design process of this system can be divided into six steps which includes selecting data sets, preprocessing image, tilt correction, character segmentation, feature extraction, and training BP network.

3.1   Selection of Data Sets

In this system we use two data sets. The handwritten number dataset is a MNIST dataset that contains sixty thousand training images and ten thousand test images.

In handwritten English letters dataset we select Chars 74K dataset. The dataset includes a to z and A to Z, it has a total of 52 classifications. Each image in the original dataset can be rotated in four different angle in order to expand handwritten English letters dataset. Hence the handwritten English letters dataset has a total of 14300 images, of which 10400 images can be viewed as the training images and the rest as the test images.

3.2   Image Preprocessing

We can set each pixel value in the image with 0 or 1 for the image binarization in this part. Then we convert 0 to 1 and 1 to 0 to make the image background color black and color of characters in the image white. Finally, the image can be normalized to 28 * 28 pixels. Like Figs. 2. and 3.

Fig. 2. This is a binary image. It is the original data set of image binarization.

Fig. 3. This is the final image in image preprocessing stage. After that we will extract features on the image in the feature extraction stage.

3.3 Tilt Correction

In this system we use the least squares method [7] for image tilt correction. Assume that there is a fitting straight line. We need to solve the slope of the line because of image tilt correction. The following formula is the equation of the slope. Finally, using this slope we can calculate the image rotation angle. Like Figs. 4 and 5.

$$k = \frac{\sum xy - n\overline{xy}}{\sum x_i^2 - n\overline{x}^2} \tag{8}$$

Fig. 4. This is an image before image tilt correction. We can see that the characters in the image is titled, not in same level.

Fig. 5. This is an image after image tilt correction. The characters in the image is in same level.

Fig. 6. This is an image. There are five characters in this image. So we need to separate the five characters.

3.4 Character Segmentation

Character segmentation algorithm ignores connections between characters, each character of the default image has interval. Character segmentation algorithm is as follows:

- Calculating all the column of pixels of an image, and storing it in a one-dimensional line matrix.
- In the one-dimensional line matrix, if a column of values is not zero, it means that this column is part of the character.
- Hence, we think that all columns of the continuous part with non-zero value can form a complete character (Fig. 7).

Fig. 7. It includes five images after using character segmentation method to process Fig. 6. These images should use image preprocessing method before feature extraction.

3.5 Feature Extraction

In this system we combine structure features with statistical features. A total of forty-one features were extracted, including sixteen pixel distribution features, eight intersection features, nine character contour features and eight projection features. The following is feature extraction method that is used in this system.

3.5.1 Pixel Distribution Features

The pixel distribution feature method is to divide an image of 28 * 28 pixel into sixteen average regions. Then we count all pixel values of "1" in each region. Finally, we can obtain all pixel value of sixteen regions as pixel distribution features.

3.5.2 Intersection Features

Intersection feature means statistics of pixel value which is one in horizontal, vertical and diagonal direction in an image. Among them, setting horizontal and vertical direction as the image of 11, 14 and 17 row and column. Adding two diagonals, it has a total of eight features.

3.5.3   Character Contour Features

In the extraction of character contour feature, we scan an image from left and right. We stop scanning and record column value at the first time when we met the pixel value of a pixel point is "1".

In this way, we have two data according to the scanning direction. Using the two data to gain other dataset. The maximum and the minimum of these dataset are used as the character contour features. It has a total of nine features.

3.5.4   Projection Features

Projection feature extraction method is to divide the image of 28 * 28 pixel into four regions on average, then we project the part of character into border of the image and record the pixel value of 1 with the corresponding row or column of the border of the image.

Since an image can be divided into four areas, the boundary of the image can be divided into eight pieces. Finally, we can extract eight projection features.

3.6   BP Network

We use the neural network toolbox of MATLAB software for the training of BP network. Using supervised learning style, we set a corresponding label for each image. A total of 62 classification include Numbers and English letters. Hence we can use the binary number six to stand for one classification. For example, we use "000000" to stand for the number "0".

Finally, we save the data after BP network training so that we can deal with test data when we test BP network in the same way.

3.7   Summary of This Chapter

According to the design and implementation of the above method, we have almost completed the system. We can use the system to carry on the handwritten character recognition.

4   Experimental Analysis

In this system, according to different training samples by BP network, we classify the experimental analysis into three cases.

4.1   First Case

In this case, we only use handwritten numbers data set when training BP network.

Table 1 shows recognition result of handwritten numbers. It shows that the recognition rate of handwritten number is from 0 to 9. The recognition rate of every classification can reach more than 90 %.

Table 1. Handwritten numbers recognition result

Classification	Correct number	Total	Recognition rate
0	958	980	97.76 %
1	1090	1135	96.04 %
2	986	1032	95.54 %
3	962	1010	95.25 %
4	926	982	94.30 %
5	846	892	94.84 %
6	926	958	96.66 %
7	973	1028	94.65 %
8	892	974	91.58 %
9	924	1009	91.58 %

4.2 Second Case

Only using handwritten English letters data set by BP network in this case, we show a part of recognition results (Fig. 8).

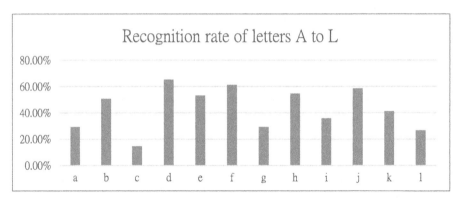

Fig. 8. It indicates handwritten English letters recognition results. We show recognition rate for letter A to L and letter a to l. The recognition results are not accurate. The recognition rate is very low for handwritten English letters.

4.3 Third Case

In this case, we put together the two different data sets to train via BP network. The following display recognition result of number from 0 to 9.

Table 2 shows the recognition result in this case when putting together the two different data set to train via BP network. By comparison with the first case, we can find the number from 0 to 9 recognition rate reduced.

Table 2. Recognition result of handwritten numbers from 0 to 9

Classification	Correct number	Total	Recognition rate
0	958	980	96.33 %
1	1090	1135	93.30 %
2	986	1032	92.54 %
3	962	1010	93.37 %
4	926	982	91.14 %
5	846	892	93.05 %
6	926	958	94.26 %
7	973	1028	92.90 %
8	892	974	82.27 %
9	924	1009	89.59 %

4.4 Results Analysis

From the above tests, it shows that the system for handwritten number recognition is available. Recognition rate is on average 94 % for handwritten number alone. However, for the recognition result of handwritten English letters is not good, so that it also affected the digital recognition rate in the Third case.

5 Conclusion

We have designed the system for recognizing the handwritten numbers and English letters. The recognition rate of handwritten numbers and English letters can reach 94 % and 44 % respectively. If the two different training data set are mixed together, the recognition rate can reach 78 %. From the above, we can conclude that the handwriting recognition of alphabet is not ideal, but for handwritten numeral recognition performance is very good. One side, there are not enough samples on the handwritten English letters. The other side, we need improve the algorithm especially on Feature extraction. In the future, we make them better by using deep learning algorithm and increasing the samples.

Acknowledgment. This work was supported by Natural Science Foundation of Fujian province of China (No. 2016J01325, No. 2015J05015), High-level Talents Foundation of Xiamen University of Technology (No. YKJ14014R and No. YKJ12025R), Science Planning Foundation of Xiamen City of China (No. 3502Z20130041 and No. 3502Z20133033), Scientific Research Fund Project of Talents of Hefei University (No. 15RC07), Natural Science Research Project of Universities of Anhui Province (No. KJ2015A206), Open Project Foundation of Intelligent Information Processing Key Laboratory of Shanxi Province (No. 2014001), Foundation of National Health and Family Planning Commission of the People's Republic of China (No. WKJ-FJ-35), Foreign Cooperation and Communication of Science Foundation of Xiamen University of Technology (No. E201401000). The corresponding author was supported by CSC .

References

1. Patel, M.S., Reddy, S.L., Naik, A.J.: An efficient way of handwritten english word recognition. In: Satapathy, S.C., Biswal, B.N., Udgata, S.K., Mandal, J.K. (eds.) Proc. of the 3rd Int. Conf. on Front. of Intell. Comput. (FICTA) 2014. AISC, vol. 328, pp. 563–572. Springer, Heidelberg (2015)
2. Liu, C.-L., Yin, F., Wang, D.-H., Wang, Q.-F.: Online and offline handwritten Chinese character recognition: benchmarking on new databases. Pattern Recogn. **46**, 155–162 (2013)
3. Wang, H., He, C.: Dynamical behavior of a cellular neural network. Acta Physica Sinica **52**(10), 2409–2414 (2003)
4. Wei, Y., Xie, J., Wu, Q.: Research on license plate character recognition based on improved LM-BP. Neural Netw. **35**(2), 48–57 (2016)
5. Deng, K.: An Automatic Marking System Based on Handwritten Character Recognition. Taiyuan University of Technology (2015)
6. Ma, C., Yu, M.: Analysis and extraction of structural features of off-line handwritten characters based on principal curves algorithm. Comput. Eng. Appl. **49**(3), 202–206 (2013)
7. Zhao, X.: Research on Recognition Method of Handwritten Numbers and English Characters. Northeast Normal University (2010)
8. Fujisawa, H.: Forty years of research in character and document recognition–an industrial perspective. Pattern Recogn. **41**(8), 2446–2453 (2008)

Method for Noises Removel Based on PDE

Baoping Xiong[1,2,3,4], Zhenhua Gan[3,4], Fumin Zou[4], Yuemin Gao[2],
and Min Du[2(✉)]

[1] Department of Mathematics and Physics, Fujian University of Technology,
No. 3 Xueyuan Road, University Town, Minhou, Fuzhou, Fujian, China
xiongbp@qq.com
[2] College of Physics and Information Engineering, Fuzhou University,
No. 2 Xueyuan Road, University Town, Minhou, Fuzhou, Fujian, China
fzugym@yahoo.com.cn, dm_dj90@163.com
[3] College of Electrical Engineering and Automation, Fuzhou University,
No. 2 Xueyuan Road, University Town, Minhou, Fuzhou, Fujian, China
[4] Key Lab of Automotive Electronics and Electric Drive Technology
of Fujian Province, Fujian University of Technology, No. 3 Xueyuan Road,
University Town, Minhou, Fuzhou, Fujian, China
{ganzh, fmzou}@fjut.edu.cn

Abstract. Among various kinds of image denoising methods, the Perona–Malik model is a representative Partial Differential Equation (PDE) based algorithm which effectively removes the noise as well as having edge enhancement simultaneously through anisotropic diffusion controlled by the diffusion coefficient. However, Partial Differential Equations (PDE) is good at removeling Gaussian noises, but it is not an ideal method to deal with salt-and-pepper noise. To realize less diffusion in the texture region and more smooth in flat region while implementing image denoising, this paper propose an improved Perona–Malik model based on new diffusion function which change with the number of iterations. The improved algorithm is applied on numerical simulation and practical images, and the quantitative analyzing results prove that the modified anisotropic diffusion model can preserve textures effectively while ruling out the noise, meanwhile, the PSNR are increased obviously.

Keywords: PDE · Noises removal · Image smoothing

1 Introduction

Image denoising is very important for image segmentation, edge detection and feature extraction [1–6]. In many practical cases, due to the high frequency characteristics and useful detailed information of noise, it is difficult to reserve edge or texture information efficiently while ruling out image noise for most of the image denoising methods [7, 8]. The basic principles of the denoising algorithms can be summarized to less blur in the texture region and more smooth in flat region [9]. According to the principles, the Partial Differential Equation based (PDE-based) image denoising methods were developed and applied in computer visualization broadly. Among them, the representative one is the nonlinear diffusion approach named P–M model proposed by Perona and Malik

© Springer International Publishing AG 2017
J. Pan et al. (eds.), *Intelligent Data Analysis and Applications*, Advances in Intelligent Systems and Computing 535, DOI 10.1007/978-3-319-48499-0_19

[10–12]. Through anisotropic diffusion controlled by the diffusion coefficient which is small while the gradient of image is large, the P–M model can effectively remove the noise as well as contain edge enhancement simultaneously. However, PDE-based denoising algorithm has inherent draw backs and then various modified approaches were proposed. Shih et al. [13] figured out that the nonlinear diffusion model was weak in removing salt-and-pepper noise and they proposed a convection–diffusion filter by adding a convection term in the modified diffusion equation as a physical interpretation for denoising. In this paper, we propose an improved Perona-Malik model based on new diffusion function which change with the number of iterations, successfully used for image denoising and detailed information preserving together. In the following section, the derivation process of the P–M model is firstly introduced and its instability is analyzed. In Sect. 3, the modified P–M model based on new diffusion function which change with the number of iterations (improved P–M model) is developed. Then improved model is employed in Len image to verify its virtues of decreasing image noise while preserving the detailed information effectively. Furthermore, quantitative comparison proves that the improved model can improve the performance of texture preserving with an ideal stability. Finally, the conclusion is given at the last section.

2   P-M Model

In 1990, the anisotropic diffusion first introduced by Perona & Malik in image processing [10]. The function is nonlinear by which smoothing is only performed in low gradient areas (homogeneous areas). Thus allowing noise blurring with edge preserving:

$$\begin{cases} \frac{\partial u}{\partial t} = div[g(|\nabla u|)\nabla u] \\ u(x,y,0) = \mu_0(x,y)t \in (0,T) \end{cases}, \tag{1}$$

Where u(x, y) is a grayscale image, represented by a function of $\Omega \subset R^2 \to R$ that associates to a pixel $(x, y) \in \Omega$ its gray level u(x, y); Ω is the support of the image. $u_0(x, y)$ is defined as a noisy version of u(x, y). g is a decreasing positive function satisfying

$$g(0) = 1$$
$$\lim_{x \to +\infty} g(x) = 0,$$

called "diffusion function", which allows to define the strength of the smoothing process for each gradient norm value. The aim of this equation is to apply a diffusion process inside the homogeneous regions, where $|\nabla u|$ is small, while the diffusion is stopped near the boundaries or edges, where there is a large grey level difference between neighboring pixels and, therefore, a large gradient. Perona & Malik defined g as following:

$$g(|\nabla u|) = \frac{1}{1 + (|\nabla u|/k)^2} \tag{2}$$

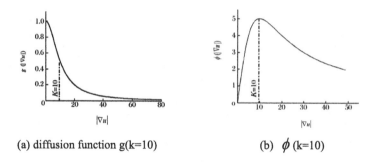

(a) diffusion function g(k=10)   (b) ϕ (k=10)

Fig. 1. g and ϕ's figure

Where k is a constant value. In order to know the function of k in the diffusion processing, a function ϕ is defined as following:

$$\phi(|\nabla u|) = \frac{|\nabla u|}{1 + (\nabla u/k)^2} \tag{3}$$

Figure 1(b) show that the diffusion is quick when $|\Delta u| \approx k$ as there is a noises region, While $|\Delta u| \gg k$ or $|\Delta u| \ll k$ the diffusion is very slow as there is smoothing regions or edge. But gradient of salt-and-pepper noise is bigger than k, so it is not easy to removel salt-and-pepper.

3   Problem and Improvement

There are some problems for the P-M diffusion function. The uniqueness and stability of the equation's solution can't be assured [12]. When some salt-and-pepper noise in the image lead to large gradient, the method can't removel the noises from the image, because the diffusion might blur the image when the k is too large, and that the diffusion would be very slow and the large gradient noise may not be removed when the k is very small.

In order to solve this problem, a new diffusion equation change with time is advanced. The diffusion equation is defined as following:

$$\begin{cases} g(|\nabla u|, t) = \cos(\frac{\pi^* |\nabla u|}{255})^{t \bmod 10} + 1 & t <= 20 \\ g(|\nabla u|, t) = \cos(\frac{\pi^* |\nabla u|}{255})^{10} & t > 20 \end{cases} \tag{4}$$

So function ϕ is defined as following:

$$\begin{cases} \phi(|\nabla u|, t) = |\nabla u|^* (\cos(\frac{\pi^* |\nabla u|}{255})^{t \bmod 10} + 1) & t <= 20 \\ \phi(|\nabla u|, t) = |\nabla u|^* \cos(\frac{\pi^* |\nabla u|}{255})^{10} & t > 20 \end{cases} \tag{5}$$

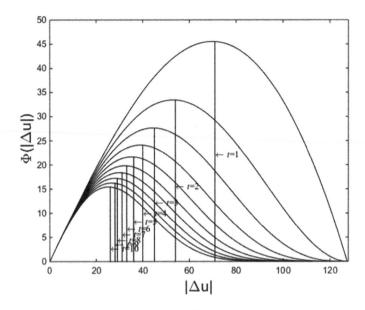

Fig. 2. Equation (3) at different time's figure

Figure 2 shows the Eq. (5) at different time's figure. We can see that the method can smooth the large gradient noises in the first place when the value of t is very small. And then the max value of $\phi(|\nabla u|, t)$'s $|\nabla u|$ is decreasing shapely as the time passes by, which decreases the blurring of image's edge as well as the noises and signal's intensity with the latter dropping more quickly. At last, the max value of $\phi(|\nabla u|, t)$'s $|\nabla u|$ keeps steadily. So the new method can removel salt-and-pepper noises while preserving edge better.

4   Implement and Result

To verify the feasibility of the modified P–M method, Lena image is employed to compare the denoising results. The bit depth of the employed gray-levele images are 512*512 and 8-bits. Figure 3(a) is the original image; Fig. 3(b) is the original image containing Gauss-type noise with the mean value of zero and standard deviation of $\sigma = 1$ and salt-and-pepper noise with density of 0.01; Fig. 3(c) is P-M filter with constant parameter k = 20, number of iterations 100 and $\lambda = 1$'s result, Fig. 3(d) is improved P-M filter with number of iterations 100 and $\lambda = 1$'s result; From c and d can't see any difference between them, but the Peak Signal to Noise Ratio (PSNR) of c and d are 26.1 and 27.4, and the mean Structural Similarity (MSSIM) between the original image and c, d are 0.7714 and 0.7735. So from PSNR and MSSIM we know that d is better than c;

To obtain an objective evaluation of the improved model, Table 1 shows the value of PSNR of Lena image with different number of iterations. From the statistical results in Table 1, all the two models perform well in noise restraining and make the noised

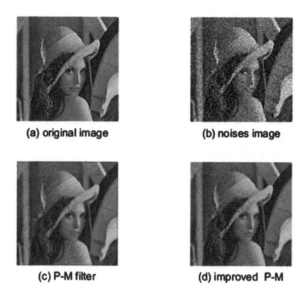

(a) original image   (b) noises image

(c) P-M filter   (d) improved P-M

Fig. 3. Different filter's result

Table 1. P-M and Improved P-M's signal-to-noise

Number of iterations	10	20	50	100	150
P-M	23.25	26.19	27.48	26.18	25.72
Improved P-M	27.41	27.40	27.40	27.42	27.43

image smoother. Improved P-M's PSNR change little With the growth of the number of iterations, the P–M model's PSNR, First, increases and then decreases. But in comparison, improved P-M's PSNR is larger than the P–M model, so the texture and detailed information are much better preserved while the noise is decreased effectively than the P-M filter model;

5   Conclusions

In this paper, the draw backs of the P–M model upon its partial differential equation have been analyzed and an improved denoising hybrid model has been proposed to avoid these drawbacks. The improved algorithm is applied on numerical simulation and practical images. The quantitative analyzing results prove that the new modified anisotropic diffusion model can preserve textures effectively while ruling out the salt-and-pepper noise. However, the iteration times decrease sharply. So in practical cases, we should take into account these factors to give suitable value to them, and sometimes, the experience is also useful to decide the close value of these parameters to get ideal results by less iteration times.

Acknowledgments. This work is partially supported by National Natural Foundation Project (61304199), The Ministry of science and technology projects for Hong Kong and Maco (2012DFM30040), Major projects in Fujian Province (2013HZ0002-1,2014YZ0001).

References

1. Bertalmio, M., Caselles, V., Pardo, A.: Movie denoising by average of warped lines. IEEE Trans. Image Process. **16**, 2333–2347 (2007)
2. Schulte, S., Huysmans, B., Pižurica, A., Kerre, E.E., Philips, W.: A new fuzzy-based wavelet shrinkage image denoising technique. In: Blanc-Talon, J., Philips, W., Popescu, D., Scheunders, P. (eds.) ACIVS 2006. LNCS, vol. 4179, pp. 12–23. Springer, Heidelberg (2006). doi:10.1007/11864349_2
3. Ehsan, N.: PDEs-based method for image enhancement. Appl. Math. Sci. **2**, 981–993 (2008)
4. Pang, Y., Yuan, Y., Wang, K.: Learning optimal spatial filters by discriminant analysis for brain–computer-interface. Neurocomputing **77**, 20–27 (2012)
5. Pang, Y., Hao, Q., Yuan, Y., Hu, T., Cai, R., Zhang, L.: Summarizing tourist destinations by mining user-generated travelogues and photos. Comput. Vis. Image Underst. **115**, 352–363 (2011)
6. Pang, Y., Yuan, Y., Li, X., Pan, J.: Efficient HOG human detection. Sig. Process. **91**, 773–781 (2011)
7. Arivazhagan, S., Deivalakshmi, S., Kannan, K.: Performance analysis of image denoising system for different levels of wavelet decomposition. Int. J. Imaging Sci. Eng. **1**, 104–107 (2007)
8. Buades, A., Coll, B., Morel, J.M.: A review of image denoising algorithms. Multiscale Model Simul. **4**, 490–530 (2005). [11] Chinna Rao, B., Madhavi Latha, M.: Analysis of multi resolution image denoising scheme using fractal transform. Int. J. Multimedia Appl. **2**, 63–74 (2010)
9. Catté, F., Lions, P.-L., Morel, J.-M., Coll, T.: Image selective smoothing and edge detection by nonlinear diffusion. Soc. Ind. Appl. Math. J. Numer. Anal. **29**, 182–193 (1992)
10. Perona, P., Malik, J.: Scale space and edge detection using anisotropic diffusion. IEEE Trans. Pattern Anal. Mach. Intell. **12**, 629–639 (1990)
11. Whitaker, R., Pizer, S.M.: A multiscale approach to nonuniform diffusion. CVGIP: Image Underst. **57**, 111–120 (1993)
12. Fischl, B., Schwartz, E.: Adaptive nonlocal filtering: a fast alternative to anisotropic diffusion for image enhancement. IEEE Trans. Pattern Anal. Mach. Intell. **21**, 42–49 (1999)
13. Shih, Y., Rei, C., Wang, H.: A novel PDE based image restoration: convection diffusion equation for image denoising. J. Comput. Appl. Math. **231**, 771–779 (2009)

Innovative E-learning
and Applications

Application of E-Learning in Teaching of English as a Second Language

Pei-Wei Tsai[1,2], Pei-Shu Tsai[3(✉)], Powen Ku[4], Vaci Istanda[5],
and Tarek Gabe[6]

[1] College of Information Science and Engineering,
Fujian University of Technology, Fuzhou 350118, Fujian Province, China
peri.tsai@gmail.com
[2] Fujian Provincial Key Laboratory of Big Data Mining and Applications,
Fujian University of Technology, Fuzhou 350118, Fujian Province, China
[3] Graduate Institute of Translation and Interpretation,
National Changhua University of Education, Changhua 500, Taiwan
pst@cc.ncue.edu.tw
[4] Graduate Institute of Sports and Health,
National Changhua University of Education, Changhua 500, Taiwan
powen@cc.ncue.edu.tw
[5] Yilan County Indigenous Peoples Affairs Office, Yilan 26051, Taiwan
vaci@mail.e-land.gov.tw
[6] Faculty of Computers and Informatics, Suez Canal University, Ismailia, Egypt
Tmgaber@gmail.com

Abstract. The present study adopted a differentiated instruction method to non-English majors at a university. An e-learning system was used as a support system for the students to check homework answers and to upload assignments. In addition, an assignment was distributed after a midterm examination to encourage the undergraduates to practice with online materials for learning outside the classroom. Students' feedback on this teaching method was performed on the e-learning system, which delivered a pre-test and a post-test questionnaire before and at the end of the period during which a differentiated activity sheet was used as an incentive for autonomous self-learning. The results showed that compared with a traditional method in which students received assignments of the same difficulty, the differentiated method, together with supports from the e-learning system, enhanced motivation for students to pursue better scores in English achievement tests.

Keywords: E-learning · Differentiated instruction · English as a second language

1 Introduction

A differentiated instruction method has drawn more and more attention to educators in the last decade [1, 2, 3]. The present study argues that for courses offered at a university level, a differentiated teaching method is in urgent need, because at university, students come from dissimilar economical and geographical backgrounds. However, even in the

© Springer International Publishing AG 2017
J. Pan et al. (eds.), *Intelligent Data Analysis and Applications*, Advances in Intelligent Systems and Computing 535, DOI 10.1007/978-3-319-48499-0_20

US, where differentiated instructions have been promoted for more than ten years since the No Child Left Behind Act announced in 2001 (NCLB) [4], not many university classes actually applied such teaching belief and method. Therefore, this present study was initiated to apply a differentiated instruction method, together with an online e-learning platform, to two different groups of non-English majors (geography vs. electrical engineering) taking the course of Freshman English as a demonstration of how differentiated instructions can be executed in undergraduate classrooms at university.

2 Related Works

In practice, a common situation for implementing a differentiated instruction method was to vary the difficulties of the teaching materials for each student according to individual learner's own learning style and preference [5]. Compared with a traditional teaching method that required every student to learn from the same textbook and be examined on the same basis, this method encouraged each student to learn according to their own pace. Reports have shown that one of the major factors for the success of differentiated instructions was the support of a small-sized classroom, extra time and care, and multiple resources for the teacher to take care of individual's needs [6, 7]. However, not every classroom in practical situation could afford such immense and expensive request from the teacher and from administration units such as the school.

The spirit of differentiated instructions was for the lecturer to provide various methods and support systems for each of the students to succeed [1]. Therefore, in the present study, we proposed that the teacher did not necessarily need to prepare 50 different textbooks for 50 different students. Instead, the teacher could vary the ways that the students hand in assignments, such as distributing differentiated homework.

In order to acknowledge the uniqueness of each student and to encourage every student in the classroom to keep up with learning English, which was a required subject but not of the main interest of students majoring in either geography or electrical engineering, the present study assigned differentiated studying activities for students taking a required course of Freshman English, together with materials provided on an on-line e-learning platform, as an incentive for students to initiate extracurricular learning by themselves.

Because one of the critical elements of a differentiated instruction was to vary the available options with different difficulties and learning styles instead of varying the learner's goals or even lowering the advanced learners' performance expectations [5], we decided to create a variety of options approachable to the students, so that the students could choose the best learning material and strategies fitting their personal needs. This was where e-learning could facilitate the teacher in teaching, with which the students could choose the difficulty level of the learning material that suits his or her interest.

3   Methods

3.1   Participants

A total of 91 Freshman English students enrolled in National Changhua University of Education were recruited in this study. Among them, 88 students volunteered and completed the study's questionnaires. The overall response rates were high for each class, with 98 % (20 females, 25 males) for the geography majors and 96 % (6 females, 37 males) for the electrical engineering majors.

3.2   Research Design

This was a within-subject design. A semester was divided into two phases by a mid-term examination into a pre-test and a post-test phase. In between these two phases, a studying sheet corresponding to the student's English proficiency level was assigned.

3.3   Materials

Studying Sheet. A studying sheet with differentiated extracurricular activities was designed. The students were separated into four groups based on the scores they received in a midterm examination. In accordance with the learners' performance, four types of studying sheets were designed for each group as described in Table 1.

The main idea of this differentiated studying sheet was to provide opportunities for students at different English proficiency levels to work on the missing skills that they

Table 1. Differentiated studying sheets

Studying sheet	Midterm scores (points)	Studying topic	Point earning rule
1	Below (including) 40	Look up new vocabulary and provide Chinese translation	One point per vocabulary and definition
2	41 ~ 60	Look up new vocabulary + make a sentence	One point per full sentence with translation of the new vocabulary
3	61 ~ 80	Look up a verbal phrase + make a sentence	One point per full sentence with translation of the verbal phrase
4	Above (including) 81	Record a two-minute summary about a movie	Three points per movie

need, as well as to encourage advanced learners to work on extracurricular activities that might be entertaining and intellectually challenging.

Questionnaire. One week before a midterm examination, the students were encouraged to fill out an anonymous questionnaire online to express their thoughts about the teaching and the learning experience in the English classroom that was taught with a traditional lecture method. Because the questionnaire was anonymous, the students could express their opinions to the lecturer without reservation. Also, because the questionnaire was distributed online, there was no handwriting to reveal the identity of the students.

A second anonymous questionnaire was distributed online one week before a final examination. In this questionnaire, in addition to an open question for the students to express freely about the English learning experience, there were another 31 questions measured on a 1–5 rating scale (from the least favorite or extreme disagreement to the most favorite or extreme agreement) evaluating three major categories of the course: the teaching output, the teaching procedure, and the teaching involvement.

3.4 Procedure

The duration of the study was one semester (18 weeks). The first eight weeks of the semester was taught with a traditional lecturer teaching method. On the eighth week before the midterm examination, the first questionnaire was distributed. After the midterm examination, a studying list adjusted to each student's English proficiency level was assigned to the student based on individual midterm score. During the period between the midterm and the final examinations, the students chose the amount of effort and the activities listed on the studying sheet to help them learn English outside the classroom. Another anonymous questionnaire was distributed one week before the final examination as an evaluation of the effect of the differentiated instruction. The scores of the final examination were also collected as another evaluation of the students' English achievement in the course.

The study encouraged self-directed learning, so the students were free to choose or to combine studying sheets that were higher than the level that they were assigned to if they found the task not challenging or interesting enough, but they were not allowed to change or switch to activities that were easier than their current English proficiency level. The students could earn up to twenty extra academic points to add up to their midterm examination scores, which accounted for 20 % of the total score. In other words, the study encouraged learner autonomy, and the studying activity was not mandatory.

4 Results

4.1 Questionnaire

Responses from the students are summarized in Table 2. The overall scores of the questionnaires were above 4 on a five-point rating scale for each class, which could be considered as a high average.

Table 2. Summary of the rating scores in the second questionnaire

Score summary			Evaluation aspect
EE[a]	G	Avg.	1. Teaching Output
3.98	3.93	3.96	1-1 Satisfaction
4.08	3.88	3.98	1-2 Effect
			2. Teaching Procedure
4.07	4.09	4.08	2-1 Learning evaluation
4.12	4.15	4.14	2-2 Learning atmosphere
4.22	4.34	4.28	2-3 Teaching attitude
4.12	4.09	4.11	2-4 Teaching methods and strategy/teaching contents
			3. Teaching Involvement
4.10	4.05	4.08	3-1 Teaching and teaching materials plans
4.10	4.05	4.08	3-2 Evaluation on students' needs

[a] *Note.* Abbreviations: EE = Electrical Engineering majors;
G = Geography majors; Avg. = Average score.

As shown in Table 2, over the evaluation items, both classes gave the highest rating for teaching attitude. The second highest score was given to learning atmosphere, in which the students considered the learning experience was relaxing and fun. The third highest score was teaching methods and strategy/learning contents. This included extracurricular activities that varied in difficulties, the mixed use of multimedia teaching materials, interactive games, and lecture. In other words, the introduction of differentiated instruction into the course curriculum, accompanied with the online e-learning system, had triggered the students' interest in learning that brought in positive feedback.

In addition to the rating scores, the students also provided their thoughts about the course in the open question. The written responses were in accordance with the findings in the rating averages. For example, one student wrote *"thank you teacher, for giving me the opportunity to learn."* The ther stated *"I am impressed by the way that the teacher taught in accordance with each student's individual talent."* Another student responded *"the course was interesting, relaxing, and enjoyable."*

4.2 Test Scores

The midterm and final examination scores for the two classes of Freshman English are illustrated in Fig. 1.

These scores presented in Fig. 1 represent the sum of the students' effort accumulated over the course. Figure 1 shows that after the students had studied with the assignments that were differentiated in levels of difficulty, the overall number of students failing the class (score points lower than 60) had been tremendously reduced in comparison to the number of students failing the midterm examination.

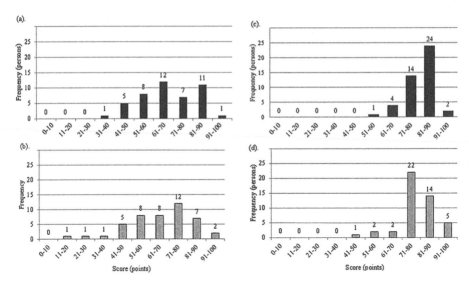

Fig. 1. Examination scores. (a). Midterm: geography majors. (b). Midterm: electrical engineering majors. (c). Final: geography majors. (d). Final: electrical engineering majors.

5 Conclusions and Discussion

The present study differed from a traditional teaching approach in that the lecturer opened an opportunity for the students to make extra effort for their scores, allowing the midterm examination scores to be added up to 100 points if the students were willing to complete all activities on the studying sheets. This modification only influenced 20 % of the students' assessment scores, which perhaps did not exert an overwhelming impact on the total score, but might have triggered the students' motivation to work harder for improving their English scores. Therefore, in the final examination, which accounted for 30 % of the total score, we observed an overall improvement across the two classes. In particular, at the midterm examination before the differentiated instructions were administrated, approximately 35 % of the class, i.e., about 16 people per class, would probably fail the class. However, at the final examination, there was only up to one student who failed, which was about 2 % of all participated students.

The results of the present study demonstrated that through activities of varying difficulties, students who were willing to make an effort for improving or exercising with English would receive some kind of rewards. Although the extra academic points that the students received only affected up to 20 % of their overall scores, the multiple options provided to the students had initiated their motivation to keep up with learning English. The effect of the modification in the teaching instructions had exerted an enormous effect on the students' learning attitude, which was ultimately reflected in their final scores.

The results of the present study also showed that the two classes of Freshman English started off with approximately similar distribution of English proficiency levels among the students, with the class of electrical engineering displaying an even wider range (Fig. 1a and b). For the geography majors, more students scored between 81 and 90, while more electrical engineering majors scored between 71 and 80. However, if comparing the distribution of the scores within each group, such as to compare the midterm and final examination scores for geography majors only (by comparing Fig. 1a with c), one could observe that the geography students who used to score between 61–70 in the midterm had improved, thus the peak in the 61–70 interval in Fig. 1c was lower than the bar of the 61–70 interval in Fig. 1a. Similarly, the majority of the electrical engineering majors who fell behind in the midterm examination had worked their way up and caught up with the majority of the others in the final examination. Therefore, in the final examination (Fig. 1d), the number of students scoring below 60 had decreased, and the number of students scoring above 60 had increased as compared with the performance in the midterm examination (Fig. 1b). A few advanced learners worked hard as well, and therefore, the overall numbers scoring higher than the interval of 71–80 also increased.

The present study showed that in general, students reacted positively toward the differentiated studying sheets, because students of all levels benefitted from the learning experience. Diligent students would be awarded with good grades if they made an effort for learning English regardless of their initial English proficiency levels. For basic learners, they could work through the course and gain good grades as long as they were willing to spend time on the assignments. For advanced learners, they could gain a sense of accomplishment because they could pick up challenging tasks on the studying sheet.

In addition to the differentiated instruction, the use of the e-learning system may have encouraged the students to invest more time in completing the assigned tasks, because they could choose the difficulty level of the materials they wanted to study, and they were free to speak up their opinions toward the teaching contents and the teaching methods because their answers to the questionnaires remained anonymous.

Different from earlier studies advocating differentiated instruction where the majority of the research was carried out in elementary [5] and secondary schools [8], the present study demonstrated that differentiated instruction was also an executable teaching method for undergraduate courses. In particular, at the current situation of the education system in Taiwan, undergraduate courses in university in fact admitted much larger number of students than elementary and secondary school classrooms did. In such large classrooms of over 45 students per class, it was inevitable that the students would come from diverse socio-economic backgrounds with different language abilities. Therefore, it has become a challenge for teachers to pay attention to and to take care of each individual's needs while maintaining the learning interests of advanced learners and assisting basic learners at the same time. The results in the present study have suggested that differentiated instructions could be a useful and practical option for university teachers to enhance the students' learning motivation for a better learning experience.

Acknowledgments. This work is supported by Ministry of Science and Technology (MOST 103-2410-H-018-002) and Fujian Provincial Education Bureau Key Project in Technology (JA15323). The authors would like to thank Yu-Jen You for her assistance in preparing the tables and collecting references for this study.

References

1. Tomlinson, C.: Differentiating instruction: why bother? Middle Ground **9**(1), 12–14 (2005)
2. Tomlinson, C., Brighton, C., Hertberg, H., Callahan, C.M., Moon, T.R., Brimijoin, K., Reynolds, T., et al.: Differentiating instruction in response to student readiness, interest, and learning profile in academically diverse classrooms: a review of literature. J. Educ. Gifted **27** (2/3), 119–145 (2003)
3. Wilson, S.: Differentiated instruction: how are design, essential questions in learning, assessment, and instruction part of it? New Engl. Read. Assoc. J. **44**(2), 68–75 (2009)
4. No Child Left Behind Act, Pub. L. No. 107–110 C.F.R. (2001)
5. Anderson, K.M.: Differentiating instruction to include all students. Preventing Sch. Fail. **51** (3), 49–53 (2007)
6. Abbati, D.G.: Differentiated instruction: understanding the personal factors and organizational conditions that facilitate differentiated instruction in elementary mathematics classrooms. Doctoral dissertation, University of California at Berkeley, CA, USA (2012)
7. King, S.: Factors associated with inclusive classroom teachers' implementation of differentiated instruction for diverse learning. Doctoral dissertation, Tennessee State University, TN, USA (2010)
8. Brimijoin, K., Marquissee, E., Tomlinson, C.: Using data to differentiate instruction. Educ. Leadersh. **60**(5), 70–73 (2003)

A Case Study on Information Literacy and the Use of Social Media for Senior Learners

You-Te Lu[1], Yi-Hsing Chang[2(✉)], and Tien-Wen Sung[3]

[1] Department of Information and Communication,
Southern Taiwan University of Science and Technology, Tainan, Taiwan
yowder@stust.edu.tw
[2] Department of Information Management, Southern Taiwan University
of Science and Technology, Tainan, Taiwan
yhchang@stust.edu.tw
[3] Fujian Provincial Key Laboratory of Big Data Mining and Applications,
College of Information Science and Engineering,
Fujian University of Technology, Fuzhou, China
twsung@fjut.edu.cn

Abstract. The social media networking has become an important portal for increasing learning opportunities and enhancing life quality. This study was attempted to explore the behavior of using social media among senior learners. The objective of this study was to find out whether the existence of the social media has had a significant impact on information literacy among the elderly in a city area in Tainan, Taiwan. Results were mixed, which led to the conclusion that, despite the adoption of social media, the information literacy has not had an all-encompassing influence on the ability of seniors to take advantage of social media. Gender issues in older adult's participation in information literacy are not significant. Media literacy enables older adult to have the skills, knowledge and understanding they need to make full use of the opportunities presented both by traditional and by new social media services. Moreover, experiences and attitudes towards using social networking vary by educational level.

Keywords: Senior learners · Social media · Information literacy

1 Introduction

New technologies can play a major part in helping older people engage with society and having a better life. Many of senior learners have enthusiastically adopted modern media to keep in touch with their relatives, old acquaintances and younger family. Studies show that the elderly use new social media tools to bridge the gap between them and the new generation as a way to re-connect with others. As the computer revolution, such as the PC and the Internet, the needs of an individual living in a changing world at a rapid pace to cope with and deal effectively with life's many challenges are essential issues. The development of e-Learning and Distance Education technologies are permitting both youths and adults to learn anywhere on their own

© Springer International Publishing AG 2017
J. Pan et al. (eds.), *Intelligent Data Analysis and Applications*, Advances in Intelligent Systems and Computing 535, DOI 10.1007/978-3-319-48499-0_21

time. Because people should validate and assess information to verify its reliability, the point is to understand the need for effective use of information.

Information literacy and lifelong learning have a strong relationship with each other that is critical to the success of every individual in the information society. Theoretically, we should pursue the goal of becoming more information literate and continuously over our lifetime. Luckily, the task of facilitating information literacy to every group of learners is more easily attained when distance and e-Learning are used. This could be a solution to the senior learners who might need more social interactions.

In fact, technology has had a tremendously impact on what it means to be social. A new social realities created by technology and what those realities mean for the individual and society is obviously happened nowadays. Nevertheless, the use of social media has both positive and negative consequences for senior learners. Although the technology has the potential to harm or enhance our social life, the key is to understand how technology affects people socially. This is the critical issue regarding technology and aging society.

2 Literature Review

2.1 Information Literacy

Information is a vital source and certainly the basic component of education, and the literacy is "the condition of being literate" (Lau 2006). The concept of "information literacy" reflects a convergence of thinking from many developments, disciplines and areas of research. With the nature of e-Learning and information communication technologies, we are likely to encounter it in everyday life. The standards of literacy are now expanded to mean more than basic literacies of reading, writing, and numeracy, but to apply to other areas including computer literacy, media literacy, and cultural literacy.

However, the two concepts – information literacy and lifelong learning are interrelated. According to Horton (2008), both of these concepts are largely self-motivated, self-directed, self-empowering, and self-actuating. For instance, the more information literate an individual becomes, the greater the self-enlightenment that occurs if practiced over an entire lifetime.

Adults come to the learning setting from many different places and with a multitude of diverse experiences. Some of the learners may feel confident in subjects like evaluating resources and searching online. On the other hand, there are voracious adults in a learning setting have never been required to do any research via Internet. Therefore, finding an interesting way to talk about information literacy of all ages is never easy. Information literacy is not just about understanding what learner read, but comprehending any data and using it. In general, to be information literate means to be able to figure out what information learners need, then to know the skills to locate, evaluate, and use this information effectively (American College and Research Libraries Association of College and Research Libraries 2000).

2.2 Social Media

The information and communication technology (ICT) has broken the barriers of space and time, enabling learners to interact anytime with more people than ever before. In fact, the use of electronic devices has overtaken traditional face-to-face communication by a wide margin. Indeed, with more powerful social technologies at our fingertips, people are more connected (Tardanico 2012). Bosch and Currin (2015) indicated that senior adults communicated with more people by using Internet.

Social media are computer-mediated tools that allow people to exchange information and enhance more interaction with others. Social media plays an important role in our daily life to keep in touch with friends, and maintain our relationships from the real world. These social-technologies include Facebook, YouTube, Line, Blogger, Twitter, Plurk, and Google+.

However, Serra (2014) indicated that social media may appear to make our life easier but at the same time it has destroyed our quality human interaction. For instance, constantly having access to more social interaction becomes too much to handle and is technically not even real. Based on the theory of Maslow's Hierarchy of Needs, social interaction should be balanced outside of the cyber world (Raacke and Bonds-Raacke 2008).

2.3 Senior Learner

Like the rest of the world, Taiwan is an ageing society. With more people in Taiwan living longer and with couples having fewer children, the aging of Taiwan's population has become increasingly obvious. According to the Department of Human Resources Development National Development Council, Taiwan became an aging society in 1993, and is forecast to become an aged society and super-aged society in 2018 and 2025 In 2060, the proportion of the total population that is elderly in Taiwan will be higher than that in the other countries, including USA, Untied Kingdom, France, Italy, and Germany (National Development Council 2014).

The increase in the number of aged people requires relevant authorities and more researchers to work on issues such as the promotion of active aging, health care, social welfare, and education. In 2006, Taiwan's Ministry of Education published a white paper entitled "Towards an aged society: Seniors education policy white paper" to provide a friendly environment for seniors with the opportunities and rights to lifelong learning.

Senior adults belong to a very specific group in terms of learning of information communication technology (ICT) that needs a special approach to lifelong learning and educational development. Senior adults use email and social media to maintain contact with family and friends both through their communication and observation of activities, such as news, photographs and discussions. As Vacek and Rybenska (2014) described, it can lead to an improved sense of meaning in senior adults' lives if they learn to use ICT well. Thus, it is necessary to help senior learners not only to know but understand what the benefits of social media are, how to use them and how they can improve their lives.

3 Research Design

3.1 Research Design and Sample

A quantitative research approach was adopted in which a survey questionnaire was distributed to senior learners within a city area in Tainan, Taiwan. Descriptive data were collected through a self-reported survey approach in which participants responded to a series of questions about themselves. Descriptive research is used to gain improved understanding of the current environment.

The target population for this paper was limited to the senior citizens of the Yong-Kang district area in Tainan, Taiwan. A total of 70 of the 78 surveys were returned to the researcher. Eight of the 70 surveys were not completed and, therefore, not useable. A total of 65 surveys were used in the analyses.

3.2 Data Analysis

The aim of this study was to explore in formation literacy and the use of social media for senior learners. Data analysis was conducted using the Statistical Package for the Social Sciences (SPSS version 21.0). Descriptive statistics were used to analyze the collected data. Specifically, frequency distributions, percentage of responses, and mean distributions were used to describe the current status of farmers' Internet adoption.

4 Results

4.1 Sample Demographics

A summary of the sample demographics is presented in Table 1. There were 65 participants to the survey. Of the respondents, 33(51 %) were male and 32(49 %) were female. Most of respondents were in 51-to-70 year range (74 %) and the majority of the respondents were educated.

What is the senior learners' experience with using social media in Taiwan? The gathered data provided an overview of the use of social media by age groups. Frequency distributions and percentage of responses described the senior learners' experience with social media in Taiwan. As showed in Table 1, approximately 46 % of the respondents had less than one-year experience on social media. Of the respondents, 54 % have had experience on social media more than 5 years. All of senior learners have had experience with using social media.

4.2 Information Literacy

As showed in Table 2, the majority of respondents had moderate literacy level on traditional literacy, but had a high information literacy level on media literacy. However, computer literacy and Internet literacy competency of participants were moderate to low.

Table 1. Descriptive statistics of participants (n = 65)

	Variable	N	Percentage(%)
Gender	Male	33	51
	Female	32	49
Age	51–60	24	37
	61–70	24	37
	71–80	14	22
	81+	3	5
Educational background	Junior High School	28	43
	Senior High school	20	31
	College/University	17	26
	Master	0	0
Years of experience with social media	Less than 1	30	46
	1–2	19	29
	2–3	14	22
	3–4	1	2
	5+	1	2

Table 2. Information literacy for participants

Measurement	Mean	SD
Traditional literacy	3.03	1.382
Computer literacy	2.79	2.023
Media literacy	4.59	1.164
Internet literacy	2.29	1.640

4.3 Social Media

Based on the responds, all of participants showed the moderate positive impact on the use of social media. As listed in Table 3, respondents had much more interaction on the use of social media.

Table 3. Use of Social Media for Participants

Measurement	Mean	SD
Entertainment	3.12	1.874
Information	3.14	1.879
Interaction	3.22	1.941
Attitude	3.14	1.892

5   Discussion

The objective of this study was to find out whether the existence of the social media has had a significant impact on information literacy among the senior learners. Results were mixed, which led to the conclusion that, despite the adoption of social media, the information literacy has not had an all-encompassing influence on the ability of seniors to take advantage of social media. Gender issues in older adult's participation in information literacy are not significant. Media literacy enables older adult to have the skills, knowledge and understanding they need to make full use of the opportunities presented both by traditional and by new social media services. Moreover, experiences and attitudes towards using social networking vary by educational level.

Research has shown that seniors may appear to benefit from better social interaction when they have more use of social media. Since the social technologies have broken the barriers of space and time enabling people to interact anytime and anywhere, we should be able to take advantage on the higher levels of social interaction via social media.

Besides, information literacy is a preparation for lifelong learning of all ages. Learners need to experience and apply information literacy at all levels of using new media. In teaching and learning terms, information literacy are expressed as important factors where seniors come to a learning situation with expectation of using social media, in turn, are affected by lifelong learning and adult educational development.

References

Association of College and Research Libraries. Information Literacy Standards for Higher Education (2000). http://www.ala.org/acrl/standards/informationliteracycompetency. Accessed 17 March 2016

Raacke, J., Bonds-Raacke, J.: MySpace and Facebook: applying the uses and gratifications theory to exploring friend-networking sites. Cyberpsychol. Behav. **11**(2), 169–174 (2008). http://www.ncbi.nlm.nih.gov/pubmed/18422409. Accessed 01 March 2016

Bosch, T., Currin, B.: Uses and gratifications of computers in South African elderly people. Media Educ. Res. J. **23**(45), 9–17 (2015). doi:10.3916/c45-2015-01

Horton, F.W.: Understanding Information Literacy: A Primer. UNESCO, Paris (2008)

Hung, J.-Y., Lu, K.-S.: Research on the healthy lifestyle model, active ageing, and loneliness of senior learners. Educ. Gerontol. **40**(5), 353–362 (2014)

Lau, J.: Guidelines on Information Literacy for Lifelong Learning. Universidad Veracruzana, Veracruz (2006)

National Development Council: Population Projections for Republic of China (Taiwan) 2014–2060. Department of Human Resources Development, Executive Yuan, Taiwan (2014)

Rybenska, K., Vacek, P.: Research of interest in ICT education among seniors. Soc. Behav. Sci. **171**, 1038–1045 (2014)

Serra, J.: Social media is destroying quality human interaction. The Thought & Expression Co., 17 September 2014. http://thoughtcatalog.com/jessica-serra/2014/09/social-media-is-destroying-quality-human-interaction/. Accessed 17 May 2016

Tardanico, S.: Is social media sabotaging real communication? 30 April 2012. http://www.forbes.com/sites/susantardanico/2012/04/30/is-social-media-sabotaging-real-communication/#73d848344fd8. Accessed 17 May 2016

Establishing a Game-Based Learning Cloud

Yi-Hsing Chang[(✉)], Jheng-Yu Chen, Rong-Jyue Fang,
and You-Te Lu

Department of Information Management,
Southern Taiwan University of Science and Technology, Tainan County, Taiwan
{yhchang, ma090110, rxf26, yowder}@stust.edu.tw

Abstract. The objective of this study was to develop a game-based learning cloud. The design concepts are as follows: first, a modular object-oriented dynamic learning environment was used as the basic platform, and virtualisation technology was employed to build a private cloud; second, games and related modules were developed on Google App Engine to enhance the overall system performance of the learning cloud; and third, a game-based learning approach was adopted to stimulate the curiosity of students, leading to active learning and enhanced student learning outcomes. After the system was completed, third-year students from the Department of Information Management who were taking Information and Communication Security as an elective course at our university, 85 students from 2 classes, were recruited for an experiment. System analysis was performed using the data obtained from the questionnaire. Finally, the experiment results were discussed, and a conclusion and recommendations for future studies were offered.

Keywords: Cloud learning · Game-based learning · Hybrid cloud · Moodle · Google App Engine

1 Introduction

With the availability of cloud computing technology, educators can currently access additional teaching materials for helping learners gain knowledge. In addition, the pay-as-you-go feature of cloud computing reduces investment costs for hardware devices, thus enabling learners to engage in Internet-based learning at all times. As cloud computing has become a research hotspot among modern technologies, more researchers pay attentions in the field of education. Masud and Huang [1] pointed that scholars have made a lot of researches on two aspects: cloud computing used in the field of education, and integration of network and e-learning. The former emphasized on information system application, teaching materials building, etc. The latter emphasized on construction of campus e-learning system, e-learning model on campus network, and so on. But the research applying cloud computing to elearning is not significantly reported until now. Although Masud and Huang built an elearning cloud, and made an active research and exploration for it, how to construct an efficient and lower-cost learning cloud was still not discuss in this paper.

Oblinger [2] indicated that computer games have become an indispensable part of society and the cultural environment and are particularly appealing to children and

© Springer International Publishing AG 2017
J. Pan et al. (eds.), *Intelligent Data Analysis and Applications*, Advances in Intelligent Systems and Computing 535, DOI 10.1007/978-3-319-48499-0_22

teenagers, whose computer-related activities at home mostly consist of playing computer games. Because computer games present challenges, stimulate player curiosity, and realise the fantasies of players, game-oriented teaching methods can engender student learning motivation.

Therefore, the objective of this study was to explore the performance and the pros and cons of the hybrid cloud developed in this research and to investigate the effectiveness of learning network security by using a game-based learning method. We used a modular object-oriented dynamic learning environment (Moodle) and Google App Engine (GAE) to develop a game-based learning cloud for students to learn about network security. The new learning cloud was expected to feature superior performance, higher learning efficiency, and lower computing resource consumption compared with traditional learning management system-based private clouds. With the learning cloud saved on the Moodle as the teaching platform, students engaged in learning solely by using Web browsers. Furthermore, because games demand a large amount of computing resources and network traffic, the GAE cloud computing platform was used to develop the games to reduce the consumption of local computing resources and network traffic.

2 Literature Review

2.1 Cloud Computing and Applications

Cloud computing is not a brand new technology but rather a concept. Buyya et al. [3] and Vaquero et al. [4] indicated that cloud computing is a parallel and distributed system that comprises a group of interconnected virtual computers. It is able to dynamically deploy one or more computers and allocate computing resources based on the contract between service providers and consumers. Mell and Grance [5] asserted that cloud computing is a model for enabling access to shared configurable resources that can be provisioned with minimal management costs according to user demands. These definitions indicate that cloud computing consists of a group of virtualised servers that deliver services to users through the Internet. These virtualised servers can be dynamically adjusted and users may select the pay-per-use option, in which payments are made based on the actual amount used, which facilitates optimal cost–benefit ratios.

In recent years, a large amount of studies have been conducted regarding the use of cloud computing in the field of e-learning and the results have shown the positive effects of cloud computing on e-learning. Liao et al. [6] presented a new model of collaborative e-learning using clod computing to solve the problem of providing inadequate support for individual learners during the whole learning process. Cheng and Chen [7] implemented a cloud-based article classification system used for learning English grammar that collected, analysed, classified, and gathered English-learning articles. Because a powerful computing resource was required to perform these operations, a virtualisation technology was subsequently adopted to build private clouds that were subsequently combined with social networks and adaptive testing mechanisms to provide learners with appropriate articles for elevating their English proficiency.

Although cloud computing provides users with a considerable amount of resources, the use of cloud computing for e-learning without jeopardising sensitive and private information remains a prominent problem. Therefore, the purpose of this study was to build a low-cost and efficient hybrid cloud to achieve privacy and security during cloud-based e-learning.

2.2 Game-Based Learning Theories and Applications

A number of scholars have conducted related studies on game-based learning, demonstrating the positive effects of game-based learning on learning outcomes. Callaghan et al. [8] investigated how virtual-world and video-game technologies can be used in electronic and electrical engineering courses to create a teaching environment that engendered high student involvement and participation. Sung and Hwang [9] developed a collaborative, game-based learning environment and integrated it with Mindtool, which enabled students to share and organize the knowledge they have gained during the game-playing process. The results showed that this teaching method could enhance the learning attitude and motivation of students and simultaneously elevate their academic performance and self-efficacy. Kim et al. [10] explored a game-based mobile learning model and found that the students were able to tackle a series of challenging mathematical problems without the help and guidance of their teachers.

In this study, we adopted an adventure game-type (AVG-type) design in which information hidden in stories was used as clues for guiding players, who were required to solve various puzzles to complete the game. Thus, the observation and analytical skills of players were tested.

3 System Design

3.1 System Architecture

A learning cloud based on game-based learning (LC-GBL) was introduced and the LC-GBL system architecture comprised a presentation layer, a service layer, and a data layer, as shown in Fig. 1.

(1) Presentation layer

The layer enabled the users using devices with Internet access to connect to the Moodle Web interface through Web browsers. Thus, users could engage in game-based learning merely by accessing Web browsers. They could also view their own learning status and game ranking. Teachers could also add, edit, and delete game-based learning materials solely by using Web browsers.

(2) Service layer

The hybrid cloud was developed Moodle and GAE.

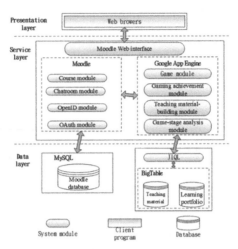

Fig. 1. System architecture

(a) Moodle: Moodle was used as the basic architecture of the LC-GBL to primarily serve the function of a game-based learning Web portal. The demographic information of the users and the content of the teaching materials regarding network security were also saved on Moodle. Virtualisation technology was employed as the basis for building the private cloud; the virtualisation platform was made by combining several physical servers and using kernel-based virtual machine virtualisation technology. Several virtual machines were subsequently installed on the platform and each virtual machine was installed with a Moodle Web site and the MySQL database. To prevent the data on the virtual machines from being inconsistent, they were synchronised to each machine by using a distributed database and a distributed file system, which ensured the accuracy of the information that users' access. To prevent the virtual machine from stalling the overall performance of the system because of overload, a load balancer was installed at the front of the system to ensure an even flow distribution to each virtual machine.

(b) The GAE was used to develop the following modules that require a substantial amount of computing resources and bandwidths. OpenID and OAuth were used to bridge communications between these modules to enhance the performance of the overall system.

(3) Data layer
The data layer contained the MySQL database and BigTable database.

- MySQL database: this database was saved in a private cloud architecture and was primarily accessed by Moodle.
- BigTable database: this database contained learning portfolio and teaching material databases. The BigTable database was a NoSQL database provided by GAE. Because it featured an SQL language and file format that differed from traditional

relational databases, JIQL was used for language conversion, a process that enabled data addition, searching, and deletion using the SQL language of relational databases.

3.2   Design of Teaching Material

The e-teaching materials were made by referencing the materials used for the course Introduction to Network Security taught at the SME e-university. The e-learning course comprised three units, which were Introduction to the Basics of Network Security, Methods of Attacks on Network Security, and Network Security Detection and Protection. Each unit consisted of three to six related courses, comprising 14 courses and 19 learning objectives. The course framework is shown in Fig. 2.

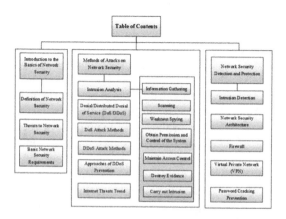

Fig. 2. The teaching materials

We began designing the content of the teaching materials for game-based learning after focused and cohesive teaching units were determined. The types of teaching materials used included text, pictures, and videos. The network security learning objectives and the content of the teaching materials for game-based learning are listed as follows:

Question: In a network environment, what are the items that should be properly protected to prevent damages from human factors and natural disasters for facilitating continued network operation and the provision of services?

Answer: In a network environment, the operating software, hardware, and operating system should be properly protected to prevent damages from human factors and natural disasters for facilitating continued network operation and the provision of services.

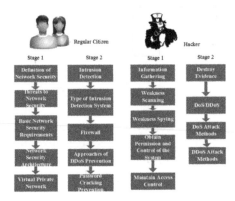

Fig. 3. Game characters and stages

3.3   Game Design

We designed the game-based learning activities by using an AVG-type game design in which each chapter was treated as a mission. Each mission comprised two scenarios and was presented in the form of shooting games. During the game, a question-and-answer format was adopted, in which students selected the answers and options by targeting their selection using an aircraft. The missions were based on AVG games and clues were provided to facilitate the completion of each game stage.

As shown in Fig. 3, two stages were designed to be completed by each game character. For a regular citizen, Stages 1 and 2 both consisted of five missions; for a hacker, Stages 1 and 2 comprised five and four missions, respectively. Each mission contained three tasks, and as the students played the game, the system recorded whether these missions had been successfully completed and a final game-stage analysis page was displayed to show the learning process of the students, enabling them to review the areas they found challenging.

4   Experiment Design and Results

4.1   Study Participants

The participants comprised third-year university students from the school's Department of Information Management who were taking Information and Communication Security as an elective course; 85 students from two classes were recruited. They were instructed to operate the system and engage in learning by using Web browsers on typical computers to play the game. Subsequently, questionnaires were distributed to the participants to obtain information regarding their experiences of using the learning system. Finally, the questionnaires were collected for statistical analysis.

4.2   Experiment Procedure

The experiment procedure was divided into three stages:

Stage 1:   During one class session, explain the study objective, how to use the system, and the duration of the experiment to the participants.

Stage 2:   Instruct the study participants to use the game-based learning system to gain network security knowledge for 2 weeks.

Stage 3:   Distribute questionnaires to the study participants on Day 14, which are to be completed in 1 h. In this stage, 85 valid questionnaires were retrieved, yielding a 100 % return rate.

4.3   Research Instrument

A questionnaire was designed for measuring the students' satisfaction with the system, and it entailed three dimensions: game-based learning, the content of the teaching materials, and operating experience. The questionnaire comprised 17 questions, and the answers were measured using a 5-point Likert scale wherein scores of 5, 4, 3, 2, and 1 represented strongly agree, agree, no opinion, disagree, and strongly disagree, respectively.

4.4   Questionnaire Analysis Results

(1) Reliability Analysis

Concerning questionnaire reliability, the Cronbach's α coefficient was used, which measured the internal consistency of the scale. A coefficient value α of 0.7 or above indicated high reliability. Reliability analysis of the various dimensions all showed an α of 0.7 or higher [11]. The α of the overall scale was 0.905, as shown in Table 1, confirming the reliability of our scale.

Table 1. Questionnaire reliability analysis

Name of subscale	Number of questions	Coefficient α
Game-based learning	7	0.896
Content of the network security teaching materials	4	0.792
Operating experience	6	0.849
Total/overall average	17	0.905

(2) Descriptive Statistical Analysis

The overall average and the overall average standard deviation of the dimension game-based learning were 3.892 and 0.787, respectively.

Regarding the dimension content of the teaching materials, the overall average and the overall average standard deviation were 4.109 and 0.707, respectively. The overall average and the overall average standard deviation for operating experience were 4.209 and 0.752, respectively.

5   Conclusion

In this study, a game-based learning cloud was developed and network security teaching materials were incorporated to enable students to learn network security knowledge that is applicable in everyday life. The cloud consisted of a hybrid cloud system architecture and developed a private cloud architecture by using Moodle. Games and related modules were integrated on the GAE by using OpenID and OAuth. Performance, experiment, and questionnaire analyses were subsequently conducted, and all the results fulfilled the study objectives. The questionnaire analysis results also indicated that the students felt that the system was easy to learn and that Web page loading speed was sufficiently fast to facilitate instant loading.

Acknowledgment. This study is supported in part by the National Science Council of Republic of China under the contract number MOST 103-2511-S-218-002.

References

1. Masud, M.A.H., Huang, X.: An e-learning system architecture based on cloud computing. World Acad. Sci. Eng. Technol. **62**, 74–78 (2012)
2. Oblinger, D.: The next generation of educational engagement. J. Interact. Media Educ. **8**, 1–18 (2004)
3. Buyya, R., Yeo, C.S., Venugopal, S., Broberg, J., Brandic, I.: Cloud computing and emerging IT platforms: vision, hype, and reality for delivering computing as the 5th utility. Future Gener. Comput. Syst. **25**(6), 599–616 (2009)
4. Vaquero, L.M., Rodero-Merino, L., Caceres, J., Lindner, M.: A break in the clouds: towards a cloud definition. ACM SIGCOMM Comput. Commun. Rev. **39**(1), 50–55 (2009)
5. Mell, P., Grance, T.: The NIST definition of cloud computing Ver. 15. National Institute of Standards and Technology: Gaithersburg, MD (2009)
6. Liao, J., Wang, M.H., Ran, W.J., Yang, S.J.H.: Collaborative cloud: a new model for e-learning. Innovations Educ. Teach. Int. **51**(3), 338–351 (2014)
7. Cheng, S.C., Chen, C.K.: Application in social network english learning based on virtual cloud technology combined with essential articles classification. J. Internet Technol. **13**(6), 989–996 (2012)
8. Callaghan, M.J., McCusker, K., Losada, J.L., Harkin, J., Wilson, S.: Using game-based learning in virtual worlds to teach electronic and electrical engineering. IEEE Trans. Ind. Inf. **9**(1), 575–584 (2013)
9. Shung, H.Y., Hwang, G.J.: A collaborative game-based learning approach to improving students' learning performance in science courses. Comput. Educ. **63**, 43–51 (2013)
10. Kim, P., Buckner, E., Kim, H., Makany, T., Taleja, N., Parikh, V.: A comparative analysis of a game-based mobile learning model in low-socioeconomic communities of India. Int. J. Educ. Dev. **32**(2), 329–340 (2012)
11. Wagner, W.E.: Using SPSS for Social Statistics and Research Methods. Pine Forge Press, Thousand Oaks (2007)

A Study on Smart Deployment for Real-Time Strategy Games

Cheng-Ta Yang[1]([✉]), Bing-Chang Chen[2], Her-Tyan Yeh[2], and Guo-Xiang Jian[1]

[1] Department of Multimedia and Entertainment Science,
Southern Taiwan University of Science and Technology, Tainan, Taiwan
{zada,ma2k0210}@stust.edu.tw
[2] Department of Information and Communication,
Southern Taiwan University of Science and Technology, Tainan, Taiwan
{bcchen,htyeh}@stust.edu.tw

Abstract. This study selected the real-time variety of strategy simulation games, i.e. real-time strategy (RTS) games, in which AI is used in tactics, unit production, reinforcement, combat, and other aspects. In RTS games, assignment and distribution of units in the battlefield are critical because they determine the survival of units and the victory or defeat of the player. In a game between a player and NPC, because the NPCs follow pre-determined algorithms, the player is able to follow the patterns of the opponent. Designing smart NPC behavior that prevents players from easily figuring out the pre-determined logic and challenges the strategies of the players is a potentially rewarding subject for research. By applying fuzzy logic to games and non-player characters' (NPC) judgment and interaction with players, players face more variations and challenges in the game, thereby impedes players from gaining insight to the game logic and increasing the fun of a game.

Keywords: Real-time strategy (RTS) · Game AI · Fuzzy system

1 Introduction

There were two categories of use of AI in NPCs in early games. One is finite state machine (FSM) [7], AI with fixed patterns allows players to understand the behavior of NPCs through moderate effort. The NPC compares the current disparity in firepower between the two players and decides to advance or retreat based on the result of the analysis. Moreover, the NPC must nonetheless stake all remaining resources in a final fight. The other category is cheating, in which NPCs directly access the player's information to generate its response. Accessing back-end information means that the NPC possesses all information in a match, including how the player develops and distributes his units, based on which to react to the situation. For instance, there is Fog of War in all RTS games to prevent players from gaining information on areas controlled by allies and from enemy territories. The NPC knows the location and resources of the player without scouting. This approach creates imbalance in the difficulty of the game and compromises playability.

J. Pan et al. (eds.), *Intelligent Data Analysis and Applications*, Advances in Intelligent Systems and Computing 535, DOI 10.1007/978-3-319-48499-0_23

AI should not be perfect because games do not need unbeatable AI. For judgment and recognition, human-like thinking can be incorporated for NPCs to think more similarly to humans, which allows a gray area in terms of determining the status of the player. Based on the determined statuses, different behavior patterns can be selected, which would improve the AI and achieve the objective of this study.

2   Game Setting

A RTS game includes resource gathering, base building, in-game technological development, and battle [1, 2, 4, 8]. This testing game focused on the reinforcement action in RTS games. Our testing focused on small-scale battle simulation, which clearly showed the distributions of units and the final results of the battle.

The core of the test was analyzing the firepower of individual units in the teams sent by the player, using fuzzy theory, to obtain overall firepower. With this information, the choice of tactics and the scale of reinforcement can be determined. Using fuzzy theory, the analysis of the firepower of players is changeable, meaning that either the NPC over- or underestimates the power of the team sent by the player, players always face different circumstances when they play the game, creating diverse possibilities and increasing playability.

2.1   Attributes Setting

Required values for all fighting units in RTS games include health, attack power and other attributes. The attributes settings for the units in this study were based on the common attributes in early strategy games.

Health. All units began with 100 points of health. When health was zero, the unit was destroyed. 100 % health meant healthy; 90 % to 80 % meant average; 70 % to 40 % meant slightly injured; 30 % to 10 % meant severely injured; and 1 % meant critical.

Attack power. The attack power settings were for calculation of damage to health points in battle. Attack power was determined by the type of weapon used by a unit. In the game, common weapons ranked from low to high attack power (handgun, sub-machine gun, rifle, shotgun, and sniper rifle, respectively).

Attack speed. After a unit fires a shot, there was delay time until the next shot. The settings were based on shots per second. To avoid too much disparity among the damages per second (DPS) of the weapons, the shots per second settings: 6 for sub-machine gun, 4 for rifles, 2 for handguns, 1 for shotgun, and 0.5 for sniper rifle.

Attack range. In the test, the distances were measured in meters. The attack ranges of units were based in reality. In reality, the effective ranges were around 50 m for handguns and shotguns, 200 to 400 m for sub-machine guns and rifles, and 800 m for sniper rifles. The effective range for handguns was 50 m.

Fighting unit. The fighting units in the game were soldiers. The difference among the soldiers was their weapon. All newly generated units in the initial stage of the game had 100 points of health and the same speed and other abilities.

Numbers. In the test for the RTS battle system, each player had 9 usable units: 2 for submachine guns, 2 for rifles, 3 for handguns, 1 for shotgun, and 1 for sniper rifle. The numbers of the types of units influenced the process of the battle. The numbers of units were assigned to each weapon based on the DPS of the weapons.

Deployment rules. We attacked enemies with the lowest health and high firepower as the top-priority tactic [6]. When encountering player's units, NPCs terminate stand-by mode and prepare for battle. There are two kinds of strategies - advancing and retreating. The decision to advance or retreat is determined by comparing the firepower of the groups of units from both sides at the moment when units from both sides meet.

2.2   Fuzzy Analysis

By adding the threat levels of different weapons at different distances to the health status of the units holding the weapons, the resulting threat levels of individual units were variable. Therefore, the analysis for this stage was divided into two parts.

The first fuzzy analysis. To determine the firepower of the enemy, we had to first obtain the threat levels and distances of weapons. In today's games, there are often multiple weapons, and there are different models of the same weapon, which presents differences in performance. We set a threat index for each weapon (DPS = attack power × attack speed), from 1 (lowest) to 10 (highest) (Fig. 1).

The effective ranges of the weapons were categorized into long, middle, and short range (Fig. 2). The basic stable effective range of a weapon was obtained when the grade of membership reached 1.

After obtaining the types of weapons, locations, and distances of players' units, the first fuzzy analysis was conducted using fuzzy rule base. Weapons had different threat levels at different distances. The analysis yielded the grade of membership of the

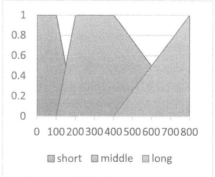

Fig. 1. Firepower membership function.   **Fig. 2.** Distance membership function.

weapons' threat levels at different distances (Fig. 3), which included categories of very strong, strong, medium, and weak.

The second fuzzy analysis. Obtaining the threat level of individual units of players. On the battlefield are units of various types in various states. There may be units with low firepower but high health or units with high firepower but low health. These two types of units have roughly the same DPS before they are destroyed. The design of the approach considered the health of the units and the threat level of the weapon they carried. As the threat levels of individual units were already obtained from the first stage of analysis, and added health levels to the unit (Fig. 4).

We obtained the grade of membership of the threat level of individual units based on fuzzy rules. The threat level of individual units influenced the priority of attack targets. As the threat levels of individual units were determined, the core process in this study was completed. Fuzzy sets still required many tests and modifications to achieve the balance and feel needed for actual gameplay. The desired effect could also be achieved through adjusting the content of the sets.

3   Experiments and Results

At the beginning of the test, players control 9 units of different types. There were fewer units with high firepower and more units with lower firepower. The player was required to coordinate his own tactics. The aim was destroying all NPC. The NPC used fuzzy logic to group units and sends them to battle player units based on the firepower of its enemies at the moment of encounter. The results of the battle were collected when the game ends with one side losing all units. The aim of the test game in this study was to create an evenly matched, unpredictable battle. When wins and losses became out of proportion, the function coefficients in the fuzzy set must be adjusted.

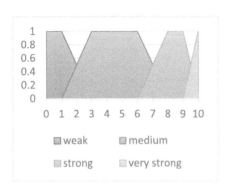

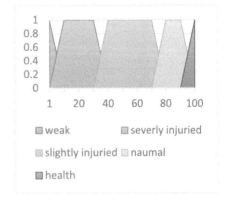

Fig. 3. Membership function of the weapons' threat levels at different distance.

Fig. 4. HP membership functions.

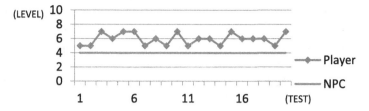

Fig. 5. The result of sniper rifle in short distance test.

3.1   Results

In short-distance tests, almost all of the player's units have higher firepower than NPCs. The lowest level, 5, occurred with units with handguns, sub-machine guns, and sniper rifles. From this, we concluded that the reason was handguns and sub-machine gun units had damages that were too low, and were in extremely inferior positions compared with sniper rifle units (Fig. 5).

In the mid-distance test, the threat levels in units with handguns, sub-machine guns, and shotguns decreased significantly. Because of compensated distance, the threat levels in units with rifles and sniper rifles increased. Units with rifles also reached peak threat level in mid-distance (Fig. 6).

Firepower disparity also appeared in the long-distance test. The threat levels of units with handguns, sub-machine guns, and shotguns decreased to nearly none. Because of the longer distance, the threat level of units with rifles slightly decreased, and the threat level of units with sniper rifles peaked (Fig. 7).

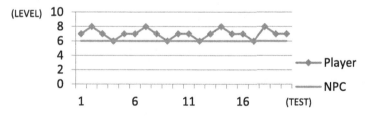

Fig. 6. The result of sniper rifle in mid distance test.

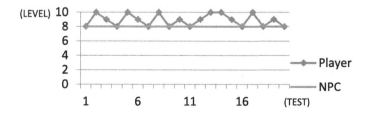

Fig. 7. The result of sniper rifle in long distance test.

3.2 Analysis

Due to limitations of space, we can only present the result of sniper rifle. During 20 tests, units from both sides began the games with full health, the player and NPC reached a tie. The initial positions of the units influenced the outcome of the battles. The deciding factor of the battles was the survival time of units with sniper rifles. The player whose unit with sniper rifle was destroyed first always lost the battle.

In testing, the NPC overestimated the firepower of its opponent frequently and sent units that were more powerful than the units of the player. This strategy may cause players to feel frustrated. The overestimation could be seen in the tests with the same types of units. The NPC usually sent reinforcement, which would cause the player to lose opportunities and lose the battle. By contrast, this situation did not occur when the NPC had superior firepower. When units from the player slowly approached units from the NPC, the NPC sent units with high firepower infrequently. Therefore, the problems were in the distances and the weapons.

4 Conclusion

In current RTS game development, use of AI has become commonplace and efficient, though limited to large-scale game productions. The results of this study can serve as useful references for small-scale teams to develop games quickly without compromising variability while considering time limit and cost.

References

1. Jónsson, B.: Representing uncertainty in rts games, Master's thesis, Reykjavík University (2012)
2. Chang, S.-H., Yang, N.-Y.: DCA: dynamic challenging level adapter for realtime strategy games. In: IEEE 15th International Conference on Computational Science and Engineering (2012)
3. Pirovano, M.: The use of Fuzzy Logic for Artificial Intelligence in Games, 7 December 2012
4. Ontanón, S., Synnaeve, G., Uriarte, A., Richoux, F., Churchill, D., Preuss, M.: A survey of real-time strategy game AI research and competition in StarCraft. IEEE Trans. Comput. Intell. AI Games (TCIAIG) **5**, 1–19 (2013)
5. Lichtenberg, V.: Fuzzy logic in computer game strategies, 29 May 2013
6. Tremblay, J., Christopher, D., Verbrugge, C.: Target selection for AI companions in FPS games. In: FDG 2014: Proceedings of the 9th International Conference on Foundations of Digital Games (2014)
7. Svensson, M., Hagelbäck, J.: Dynamic Strategy in Real-Time Strategy Games, Linnaeus University, Faculty of Technology, Department of Computer Science (2015)
8. On, C.K., Tong, C.K., Teo, J., Alfred, R., Cheng, W., Guan, T.T.: Self-evaluation of RTS troop's performance. J. Teknologi **76**(12) (2015)

A Learning Approach to Hierarchical Features for Automatic Music Composition

Michele Della Ventura[✉]

Department of Technology, Music Academy "Studio Musica", Treviso, Italy
dellaventura.michele@tin.it

Abstract. Artificial Evolution has shown great potential in the musical domain. One task in which Evolutionary techniques have shown special promise is the automatic music composition. This article describes the development of an algorithm for generating tonal melodies. The method employed does not entail any preset rule with respect to the musical grammar. It is based on a self-learning model that combines a Markov process, for the creation of concatenation rules of various sounds, with the Viterbi algorithm, for compliance with the musical syntax. The article is going to demonstrate the effectiveness of the method by means of some examples of its production and is going to indicate ways to improve the method.

Keywords: Automatic music composition · Feature learning · Hidden Markov Model · Self-learning · Viterbi algorithm

1 Introduction

Artificial Intelligence (AI) refers to the creation of information systems capable of executing tasks that are normally performed by human beings: Music Composition is one such area. With the current AI techniques, it seems difficult to create music (a melody) that might be appreciated by people [1]. Although music has a lot of features that may be formalized mathematically (such as for instance the transposition, the inversion, …) [2, 3], it also has properties that may not be so easily described from a mathematical standpoint. Such is the case of syntax of a tonal music phrase. Syntax deals with the actuals composition and it describes music on the level of its formal construct [4, 5]. Many of the studies performed in this field call for the creation of a melody neglecting the idea of form: this might work well for non-tonal music or the Jazz/Blues style, but not for a tonal melody.

This article presents an algorithm capable of generating a tonal music melody on the basis of a self-learning system, based on the idea of musical syntax.

This paper is structured as follows. Section 2 method and related work. Section 3 theory of the musical syntax. Section 4 the Process of Markov. Section 5 the algorithm of Viterbi. Section 6 methods and initial results. Section 7 conclusions.

© Springer International Publishing AG 2017
J. Pan et al. (eds.), *Intelligent Data Analysis and Applications*, Advances in Intelligent Systems and Computing 535, DOI 10.1007/978-3-319-48499-0_24

2   Method and Related Work

The possibility to create music by means of computer algorithms, has attracted a lot of interest in the last few decades [6–14]. These algorithms use different methods: Markov chains [6, 13]; knowledge-based systems [7, 9]; grammars [7, 8]; evolutionary methods divided into genetic algorithms [10] and learning systems [11–14]. Each of these studies provided important contributions to research thanks to the different perspectives that were used. However, musical syntax has never received, within the context of the tonal melody, the right amount of attention.

This article shall present an algorithm – built on the previous works – that is aimed at generating a fully-formed "tonal music melody". Though simple, the new musical idea must stem from a well-defined compositional logic that is not formalized beforehand, but that is going to be automatically and gradually built, by analyzing the harmonic structure contained in the already existing musical compositions (of tonal style and by different authors). The algorithm created for this purpose has the main task of reading from the various music pieces, the harmonic structure characterizing every single movement so as to determine the concatenation modality of the different scale degrees (musical grammar), on the basis of which the melody shall be subsequently generated. Making use of the Markov process, the algorithm will be able to improve ever more the quality of the musical ideas, by reading an ever larger number of musical works. Simultaneously, the algorithm performs a segmentation of the musical phrases subjected to analysis (musical structure), in order to identify, by comparing them, common elements that will subsequently be used for the creation of the melody. In particular, the algorithm shall identify the elements (such as for instance the cadence in its various forms (see paragraph 3)) and their position within the phrases. That is to say musical syntax is considered. This way, during the creation of the melody starting from random sounds, by using the Viterbi algorithm, it will be possible to guide the very same sounds on the basis of the concatenation probabilities acquired from the Markov process, so as to be able to obtain a correct tonal music phrase.

3   Grammar and Musical Syntax

If musical grammar consists of describing and teaching the basic elements (the notes, the pitches, the rhythm, the chord, the concatenation of the chords and so on) the syntax deals with the actual composition and it describes music on the level of its formal construct [15]. In this regard many theorists coined and then defined a specific terminology that proved extremely useful for the description and the musical analysis of a particularly large repertoire [16].

The *Motif* is the smallest syntactic unit detectable in the analysis of a composition [15]: to be identified as a "motif" a certain syntactic riff must be recurrent in its rhythmic (as a defined and characteristic sequence of durations) and/or melodic role (as a defined and characteristic sequence of intervals) [17]. The motives are structured and are made up of larger units that give birth to the phrase and to the period [16]. The phrase is a riff that contains one or more motives, spanning across several beats, completing a harmonic itinerary that leads from the tonic of a certain tonality to its dominant, on which it ends [17]. The period, made up of two (or more) phrases, is a larger syntactic riff, that from a harmonic standpoint covers a longer and more articulated route: from the tonic of a certain tonality it leads first of all to the dominant (first phrase) and then goes back from the dominant towards the tonic (second phrase), on which the period ends [17].

Musical Grammar. The objective of this study is to create a melody that should have the structure of the period and, therefore, it is important to analyze how the period is built internally.

Melody means the sequence of sounds of different pitches and durations (Fig. 1a) [16]. These sounds must belong each to a harmonic structure: i.e. harmony considers the simultaneity of several sounds (Fig. 1b) [18].

a)   b)

Fig. 1. Example of a melody (a) and example of harmonization of the melody in Fig. 1a (b).

The history of music and the analysis of the existing musical literature (related to Western music in this case) [15], permitted the definition of a well-described grammar, i.e. that set of rules needed to build the phrase and the period [18]. According to this grammar, derived from a centuries-old "tradition" of composition, the sounds that belong to a specific chord (harmony), may join (concatenation) the sounds of a subsequent chord that is connected to the preceding one [18].

Table 1 shows an example of a good and regular concatenation of the chords [18]. The first column contains the various degrees of the musical scale whereas each row indicates the degrees with which every single degree may concatenate. We must specify that there may be exceptions by way of which a certain degree resolves to a degree different than the indicated ones: actually we must not forget the creative aspect of the composer.

Table 1. Example of chord concatenation

I	II	III	IV	V	VI	VII
II	I	III	V			
III	I	II	IV	VI		
IV	II	III	V			
V	I	III	IV	VI		
VI	II	IV	V			
VII	I	V				

Musical Syntax. Musical syntax describes music on the level of its formal construct [29]: in other words, it takes into consideration not only individual chords, but the harmonic sequences as well. Even in this case, the analysis of the musical literature permitted the identification of certain characteristic elements that are always present in all the musical phrases and periods as for example:

- the "cadence", a particular harmonic sequence that assumes a very important role in the phrasing [16,17]: it actually is used in connection with musical points of interest such as conclusions, suspensions, exclamations and so on; the most common cadences are the perfect cadence, the imperfect cadence, the interrupted (deceptive) cadence and the plagal cadence (Fig. 2) [18];

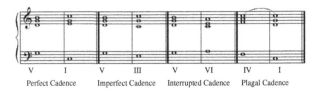

| V | I | V | III | V | VI | IV | I |
| Perfect Cadence | | Imperfect Cadence | | Interrupted Cadence | | Plagal Cadence | |

Fig. 2. Musical cadences.

- the stylistic features: harmonic sequences that are realized only in those particular ways that tradition passed on (Fig. 3) [18].

| I | II | III | IV | V | III | II | I | III | I | VI | V | IV |

Fig. 3. Musical stylistic features.

The main problem of the algorithms developed for the automatic creation of a melody consists in the fact that the melody presents itself as a sequence of sounds that belong to chords (or harmonies) that seen in sequence do not observe the good and regular concatenation (see Table 1). Furthermore, in most cases the melodies do not end up having a cadence which is typical of tonal music.

The algorithm we intend to present tries to autonomously learn the rules of musical grammar using the Markov process. At the same time, the application of the Viterbi theorem shall allow orienting the sequence of sounds so as to complete the melody with a cadence.

4 Hidden Markov Model (HMM)

The Markov chains are a stochastic process, characterized by Markov properties.

It is a mathematical tool according to which the probability of a certain future event to occur depends uniquely on the current state [19–21].

The algorithm realized uses a transition matrix to learn the rules of musical grammar: the matrix represents the probability for a type of chord to resolve to another chord. The first and foremost task of the algorithm is to read musical works in MIDI format, recognize the harmonies of different scale degrees [18] and update the transition matrix. By reading an ever larger number of musical works, the algorithm will be capable of proposing ever more ear-pleasing musical ideas: because, not only do the transition probabilities change along with the reading of musical works, but also the potential *"transition states"* (see Table 1).

5 The Viterbi Algorithm

The Viterbi algorithm is generally used to find the most likely sequence of hidden states (known as the *Viterbi path*) in a sequence of events observed in a Markov process [22], i.e. a process in which the probability to be in a state in a certain instant only depends on the state in the preceding instant.

Let us define [22, 23]:

$$\delta(i, t) = \max_{\pi(i,t)} P(\pi(i,t)|w) \qquad (5.1)$$

where $\pi(i,t)$ represents the initial part of a sequence $S1,\ldots,St$ that ends in the state i. It follows that $\delta(i, t)$ represents the probability associated to the most likely route of the first t symbols of sequence S that ends in the state i [23]. The algorithm scans all the nodes at the time t keeping track of the most likely for which there is a path with the node at the time $t-1$.

Figure 4 shows an example of harmonic sequence that starting from the I degree of the scale leads to the III degree, with an imperfect cadence, on the basis of the transitions described in Table 1. At instant $t = 1$ we hypothesize the initiation of the sequence with the I degree of the musical scale. The gray circles without an indication of the degree indicate the possibility to start from other degrees of the scale. At every subsequent instant ($t = 2, t = 3, \ldots$) we indicate the degrees with which the preceding degree may concatenate on the basis of the transition probabilities obtained according to the Markov process. The gray circles without an indication of the degree indicate the poossibility of concatenation with other degrees, not yet identified by the algorithm during the reading of the scores (see paragraph 4).

Based on this diagram, it is easy to infer how the chord sequences may change on the basis of a reading of an ever larger number of scores by the algorithm: this way new *"transition states"* may be identified and, therefore, there are more different possibilities of composition.

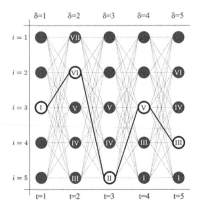

Fig. 4. Trellis diagram of a harmonic sequence (I-IV-II-V-III).

6 The Obtained Results

The algorithm developed has the objective of proposing a new tonal music idea, as a starting point for a new piece. As such, the new idea must have the distinctive feature of a musical phrase, i.e. it cannot contain modulations (passage from a tonality to another tonality), the first sound must belong to the chord of the I degree of the musical scale and it must end up with a cadence. As far as the rhythmic structure is concerned, we have deliberately not taken into consideration this element, although it is typical for music, in order to give the composer (whether a student or a professional) the possibility to refine it by inserting melodic figurations. Therefore, only one sound shall correspond to every single movement, giving the melody a homorhythmic character (every movement shall have the same duration). The only parameters required as input for the processing are the tonality and the number of movements. The first step of the algorithm was to create a transition matrix (Fig. 5) that would define the musical grammar, by reading about 300 scores by different authors and of different historical periods.

	I	II	III	IV	V	VI	VII
I		13	6	26	17	24	14
II	40		28	1	14	9	8
III	8	33		41	3	13	2
IV	8	12	24		48	0	8
V	41	2	18	12		26	1
VI	1	10	2	51	31		5
VII	96				4		

Fig. 5. Transition matrix: for every degree we indicate the percentages of transition probability.

Simultaneously, the algorithm performed a segmentation of every read score in order to search for elements (and the position of every one of them) shared with the previous scores (Fig. 6). In the segmentation phase, the score is decomposed into musical objects having a different length and the algorithm determines, for every one of them, the length, the position and the number of compositions which contain the object

Position = Number of movement - Length of the Harmonic Sequence

Fig. 6. Identification of a conclusive cadence on compositions having different lengths.

Table 2. Musical Syntax

Harmonic Sequence	Length	Position	Number of compositions
II–V–I	3	Movements-Length	48/300
IV–V–I	3	Movements-Length	236/300
...	...	...	...

(Table 2): a musical syntax is defined. This allowed for an identification of the conclusive cadence that, as already mentioned, represents a fundamental element of a tonal music idea. No rule was preset, therefore the algorithm learns autonomously how to conclude a phrase, inserting a cadence.

Finally, the algorithm will create the music idea (in full), starting to build a sequence of random chords that will have to follow one another based on the probabilities identified in the transition matrix (Fig. 7), excluding the chords that do not succeed in concatenating themselves regularly. Making use of the Viterbi algorithm, for every step (or chord created) the algorithm analyzes the possibility to conclude the music idea with a certain cadence identifying the shortest route.

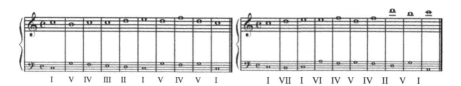

I V IV III II I V IV V I   I VII I VI IV V IV II V I

Fig. 7. Example of melodies created on the basis of a harmonic sequence.

Below there are 2 examples of melodies created considering a music idea of 10 movements.

7   Conclusions

This article presented an algorithm to automatically generate a tonal melody. The method proposed allows observing both the musical grammar related to the sound sequence and the syntax of the musical phrase: the possibility to create a fully-formed tonal music phrase, with a conclusion sanctioned by a cadence. It is easy to see how the development of algorithms for the composition of musical melodies may radically change the music composition process. However, this may not be seen as yet another case in which the computer can replace a person in a sophisticated activity, but rather as a means of support to the composer: just like a pedagogical expert system does not supersede the role of human teachers, but enables new ways to do their work. Given that music is one of the arts that has a stronger mathematical background, it is not surprising that most of the debate on whether machines can make original and creative works has centered on this sub-field of computational creativity. Hybridization of different techniques and the use of high performance computing might bring about new realms of (computer) creativity.

A future study might look into the analysis of the rhythm of music compositions, written in different forms, in order to give the melody a typical and well-defined appearance.

References

1. Miranda, E.R., Biles, J.A.: Evolutionary Computer Music, vol. 7. Springer, London (2007)
2. Cambouropoulos, E.: How similar is similar? Musicae Scientiae. In: Discussion Forum 4B, pp. 7–24 (2009)
3. Della Ventura, M.: Evaluation of musical similarity on the symbolic level of the musical text. In: Proceedings of the 15th International Conference on Artificial Intelligence, Las Vegas, USA (2013)
4. Nielsen, R.: Musical Forms. Bongiovanni Editore (1961)
5. Baroni, M., Jacoboni, C., Dal monte, R.: The music rules. In: EDT/SidM (1999)
6. Cambouropoulos, E.: Markov Chains as an aid to computer-assisted composition. Musical Praxis 1(1), 41–52 (1994)
7. Cope, D.: Experiments in Musical Intelligence. A-R Editions Inc., WI (1996)
8. Thom, B.: An interactive improvisational music companion. In: Proceeding of the Fourth International Conference on Autonomous Agents, Barcelona, Spain, pp. 309–316 (2000)
9. Miranda, E.: Composing Music with Computers. Focal Press, Oxford (2001)
10. Rowe, R.: Machine Musicianship. MIT Press, Cambridge (2004)
11. Chan, M., Potter, J., Schubert, E.: Improving algorithmic music composition with machine learning. In: Proceedings of the 9th International Conference on Music Perception and Cognition, ICMPC (2006)
12. Lichtenwalter, R.N.: Lichtenwalter, K., Chawla, N.V.: Applying learning algorithms to music generation. In: Proceedings of the 4th Indian International Conference on Artificial Intelligence, IICAI (2009)
13. Della Ventura, M.: Automatic tonal music composition using functional harmony. In: Agarwal, N., Xu, K., Osgood, N. (eds.) SBP 2015. LNCS, vol. 9021, pp. 290–295. Springer, Heidelberg (2015). doi:10.1007/978-3-319-16268-3_32
14. Smith, J.B.L., Chew, E.: A meta-analysis of the mirex strcture segmentation task. In: ISMIR (2013)
15. Schonber, A.: Theory of Harmony. University of California Press (1983)
16. Nielsen, R.: Le forme musicali. Dongiovanni Editore (1961)
17. Bent, I.: Analysis. Macmillan Publishers Ltd., London (1980)
18. Coltro, B.: Lezioni di armonia complementare. Zanibon (1979)

19. Bengio, Y.: Markovian models for sequential data. Neural Comput. Surv. **2**, 129–162 (1999)
20. Bini, D.A., Latouche, G., Meini, B.: Numerical methods for structured Markov chains. Oxford University Press, New York (2005)
21. Kazi, N., Bhatia, S.: Various artificial intelligence techniques for automated melody generation. Int. J. Eng. Res. Technol. **2**, 1646–1652 (2013)
22. Viterbi, A.J.: Error bounds for convolutional codes and asymptotically optimum decoding algorithm. IEEE Transaction on Information Theory **13**, 260–269 (1967)
23. Viterbi, A.J.: Convolutional codes and their performance in communication systems. IEEE Trans. Commun. Technol. COM–19, 751–772 (1971)

Networks and Security

Error Control Schemes for Robust Transmission with Compressed Sensing Signals

Hsiang-Cheh Huang[1], Po-Liang Chen[1], and Feng-Cheng Chang[2(✉)]

[1] Department of Electrical Engineering,
National University of Kaohsiung, Taiwan, Republic of China
hch.nuk@gmail.com
[2] Department of Innovative Information Technology,
Tamkang University, Taiwan, Republic of China
135170@mail.tku.edu.tw

Abstract. Compressed sensing is famous for its compression performances over existing schemes in this field. We apply compressed sensing to digital images for error-controlled transmission in this paper. For assessing the compression performances, researches assume to have the error-free transmission between the encoder and the decoder. For transmitting compressed sensing signals over lossy channels, error propagation would be expected, and the ways to apply error control schemes for compressed sensing signals would be much required for guaranteed quality of reconstructed images. We propose to transmit compressed sensing signals over multiple independent channels for error-controlled transmission. By employing the correlations between the compressed sensing signals from different channels, induced errors from the lossy channels can be effectively alleviated. Simulation results have presented the reconstructed image qualities, which depict the effectiveness of the use of multi-channel transmission of compressed sensing signals.

Keywords: Compressed sensing · Error control · Multiple channel

1 Introduction

Compressed sensing is one recently developed technique in lossy data compression researches and applications [1, 2]. Lossy compression serves as an inevitable part in multimedia communications. With the widely use of smart phones or tablets, increasing numbers of multimedia contents are accumulated rapidly. These contents, mostly images, can be stored, transmitted, and shared with the use of social networking services and wireless networks, which are commonly encountered in our daily lives. Thus, how to efficiently perform data compression, in addition to the robust transmission of multimedia contents, would be much required for practical applications [3, 4]. Compressed sensing presents different perspectives in lossy data compression, and with the consideration of data transmission over channels, they lead to the new branch in researches with potential applications.

© Springer International Publishing AG 2017
J. Pan et al. (eds.), *Intelligent Data Analysis and Applications*, Advances in Intelligent Systems and Computing 535, DOI 10.1007/978-3-319-48499-0_25

In compressed sensing, it requires the sampling rate, which is far less than the Nyquist rate, with the capability of reconstructing the original signal for lossy compression. The major goal in compressed sensing researches would be the compression capability. Thus, how to effectively decode with a small amount of compressed signals would be the major challenge [5, 6]. In addition to looking for compression performances, we consider the error-controlled transmission of compressed sensing signals. In this paper, we transmit compressed signals over lossy channels to observe the effect caused by packet losses. To alleviate the quality degradation, we employ the transmission over multiple independent channels. For the better protection of compressed sensing signals, improved quality of reconstructed image can be observed for the error controlled transmission.

We briefly describe the fundamentals and notations of compressed sensing in Sect. 2, present the proposed method for transmitting compressed sensing signals with the error-controlled protection over lossy channels in Sect. 3, demonstrate the simulation results in Sect. 4, and address the conclusions in Sect. 5 as follows.

2 Fundamentals of Compressed Sensing

In compressed sensing, based on the representations in [1, 2], it is composed of the *sparsity* principle, and the *incoherence* principle.

- For the *sparsity* principle, it implies the information rate in data compression. In compressive sampling, it is expected to be much smaller than the sampling rate required, and can be represented with the proper basis Ψ, $\Psi \in C^{N \times N}$, and C means the complex number. More specifically, Ψ is the basis to reach sparsity with a k-sparse coefficient vector $\mathbf{X}$, $\mathbf{X} \in C^{N \times 1}$, with the condition that

$$\mathbf{f} = \Psi \mathbf{X}. \tag{1}$$

Here, $\mathbf{f}$ denotes the reconstruction corresponding to the original signal.

- For the *incoherence* principle, it extends the duality between time and frequency. The measurement basis Φ, $\Phi \in C^{m \times N}$, which acts like noiselet, is employed for sensing the signal $\mathbf{f}$, with the condition that

$$\mathbf{Y} = \Phi \mathbf{f}. \tag{2}$$

Here, $\mathbf{Y}$ denotes the measurement vector. We note that Eq. (2) is an underdetermined system.

With the parameter settings in [1, 2], we choose K_1 and K_2 coefficients for sparsity and incoherence, respectively.

3   Transmission of CS Signals Over Multiple Channels

For the effective delivery of compressed sensing signals, and considering the robust transmission depicted in [7–9], we employ the use of transmission over multiple lossy channels, which are mutually independent, in this paper. Figure 1 describes the block diagram of our system. At the beginning, in Fig. 1(a), the input image $\mathbf{X}$ is compressed with compressed sensing in Sect. 2, and compressed sensing signals are denoted by $\mathbf{Y}$. For the ease of separating the compressed coefficients, we choose the odd-numbered indices to form $\mathbf{C}_1$, and the even-numbered ones to form $\mathbf{C}_2$. After that, we employ the multiple description transform coding (MDTC) [7] to form the two descriptions of $\mathbf{D}_1$ and $\mathbf{D}_2$, with Eq. (3).

$$\begin{bmatrix} D_{1,k} \\ D_{2,k} \end{bmatrix} = \begin{bmatrix} r_2 \cos\theta_2 & -r_2 \sin\theta_2 \\ -r_1 \cos\theta_1 & r_1 \sin\theta_1 \end{bmatrix} \begin{bmatrix} C_{1,k} \\ C_{2,k} \end{bmatrix}, \tag{3}$$

with the conditions of $\mathbf{C}_1 = \bigcup_{\forall k} C_{1,k}$, $\mathbf{C}_2 = \bigcup_{\forall k} C_{2,k}$, $\mathbf{D}_1 = \bigcup_{\forall k} D_{1,k}$, $\mathbf{D}_2 = \bigcup_{\forall k} D_{2,k}$, and $r_1 r_2 \sin(\theta_1 - \theta_2) = 1$ to lead to the determinant of one. Next, $\mathbf{D}_1$ and $\mathbf{D}_2$ are transmitted over two independent lossy channels, with the loss probability of $p_{e,1}$ and $p_{e,2}$ for Channel 1 and Channel 2, respectively, as depicted in Fig. 1(b). Due to the induced errors, the two descriptions have become $\mathbf{D}'_1$ and $\mathbf{D}'_2$, respectively.

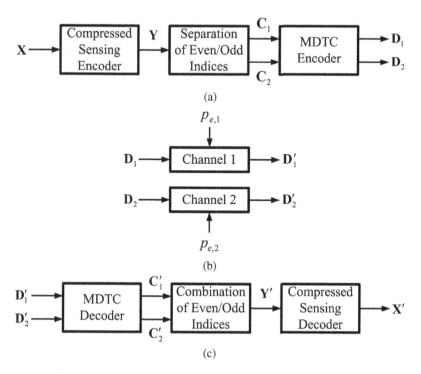

(a)

(b)

(c)

Fig. 1. Block diagrams and mathematical notations in this paper.

At the decoder, as shown in Fig. 1(c), by employing the variances as depicted in [7], and by taking the inverse operations to Eq. (3) for reconstruction, we can obtain $\mathbf{C}'_1$ and $\mathbf{C}'_2$ from received descriptions. After the combination of the even- and odd-indexed components, we can obtain $\mathbf{Y}'$. Finally, with compressed sensing, we can calculate the reconstructed image $\mathbf{X}'$.

4 Simulation Results

With the proposed method, we choose the test image cameraman, with the size of 128×128, for conducting simulations. We set $r_1 = r_2 = 1$, and choose different θ_1, θ_2, and lossy probabilities for verifications.

Figure 2 depicts the transmission of compressed sensing signals with $K_1 = 1000$ and $K_2 = 20000$, under the lossy rates of $p_{e,1} = p_{e,2} = 0.1$. For simplicity, we set $\theta_1 = -\theta_2$ or $\theta_1 - \theta_2 = \frac{\pi}{2}$ in our simulations. Reconstructed image qualities are measured by the peak signal-to-noise ratio (PSNR) and the structural similarity index (SSIM) [10], respectively. Figure 2(a) presents the results for transmitting over the single channel without the error control. Figure 2(b) to (e) display the reconstruction over two channels with different settings of θ_1 and θ_2. In Figs. 2(b), (c) and (d), we set $\theta_1 = -\theta_2$, and in Figs. 2(d) and (e), we have $\theta_1 - \theta_2 = \frac{\pi}{2}$. We can easily observe the improvements of transmission over two channels. Even though the PSNR values have slightly improved, the SSIM indices are greatly enhanced to point out the alleviation of the effects caused by the 10 % random losses from both channels. When we compare reconstructed images between Figs. 2(a) and (b), in PSNR, only the 1.29 dB improvement can be observed, and in SSIM, the enhancement of 0.39 can be watched. When we assess the two images subjectively, we may lead to the result that SSIM index may present better with the subjective point of view.

Figure 3 presents the different settings of θ_1 and θ_2 under the lossy rates between 1 % and 25 %. Again, we set $\theta_1 = -\theta_2$ or $\theta_1 - \theta_2 = \frac{\pi}{2}$ in our simulations. Figure 3(a) shows the curves with PSNR values. As expected, reconstructed image qualities become degraded with the increase of lossy rates. Without error control, or the one for single channel transmission, it generally performs the worst for lossy rates less than 18 %. Under this condition, transmission of compressed sensing with multiple channels has presented the capability for error-controlled transmission. For larger lossy rates, due to the increased number of lost packets, we can hardly find improvements with multi-channel transmission. In Fig. 3(b), we can observe similar phenomena with the SSIM curves. When the lossy rates are lower than 21 %, proposed method shows improvements over the case of single-channel transmission. Again, for larger lossy rates, received signals may be unsuitable for providing the reconstruction capability. It may lead to the degradation of reconstructed image even when error-controlled scheme is applied.

(a) Without error control
PSNR: 28.42dB SSIM: 0.22

(b) $\theta_1 = \frac{\pi}{3}$, $\theta_2 = \frac{-\pi}{3}$
PSNR: 29.71dB SSIM: 0.61

(c) $\theta_1 = \frac{\pi}{4}$, $\theta_2 = \frac{-\pi}{4}$
PSNR: 29.81dB SSIM: 0.62

(d) $\theta_1 = \frac{\pi}{6}$, $\theta_2 = \frac{-\pi}{6}$
PSNR: 28.92dB SSIM: 0.45

(e) $\theta_1 = \frac{\pi}{3}$, $\theta_2 = \frac{-\pi}{6}$
PSNR: 29.88dB SSIM: 0.62

(f) $\theta_1 = \frac{\pi}{6}$, $\theta_2 = \frac{-\pi}{3}$
PSNR: 29.76 dB SSIM: 0.62

Fig. 2. Simulations under 10 % lossy rates for test image `cameraman`.

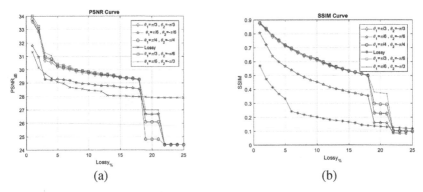

Fig. 3. Simulations of reconstructed image qualities for lossy rates between 1 % and 25 %. (a) Qualities with PSNR. (b) Qualities with SSIM.

5   Conclusions

In this paper, we have employed the use of multiple-channel transmission for the error control of compressed sensing images. With the transmission over multiple channels, and by use of the coefficient reconstruction with multiple description transform coding, proposed method has presented the effectiveness over the case with single channel transmission. However, when the lossy rates become too large, even the error controlled schemes over multiple channels may hardly provide the improvements as compared to single channel transmission. With the selection of parameters for multiple channel transmission, we have presented the effects and advantages for error-controlled transmission with compressed sensing. Other suitable combinations of parameters that may also be applicable for enhanced quality in reconstructed images may also be explored in the future.

Acknowledgement. The authors would like to thank Ministry of Science and Technology, Taiwan, R.O.C., for supporting this paper under Grants No. MOST 104-2221-E-390-012 and MOST 105-2221-E-390-022.

References

1. Romberg, J.: Imaging via compressive sampling. IEEE Sig. Proc. Mag. **25**, 14–20 (2008)
2. Candes, E.J., Wakin, M.B.: An introduction to compressive sampling. IEEE Sig. Process. Mag. **25**, 21–30 (2008)
3. Arildsen, T., Larsen, T.: Compressed sensing with linear correlation between signal and measurement noise. Sig. Process. **98**, 275–283 (2014)
4. Huang, H.C., Chang, F.C.: Error resilience for compressed sensing with multiple-channel transmission. J. Inf. Hiding Multimedia Sig. Process. **6**, 847–856 (2015)
5. Candes, E.J., Tao, T.: Decoding by linear programming. IEEE Trans. Inf. Theor. **51**, 4203–4215 (2005)

6. Chambolle, A.: An algorithm for total variation minimization and applications. J. Math. Imaging Vis. **20**, 89–97 (2004)

7. Wang, Y., Orchard, M.T., et al.: Multiple description coding using pairwise correlating transforms. IEEE Trans. Image Process. **10**, 351–366 (2001)

8. Goyal, V.K., Kovacevic, J.: Generalized multiple description coding with correlating transforms. IEEE Trans. Inf. Theor. **47**, 2199–2224 (2001)

9. Huang, H.C., Chen, P.L., Chang, F.C.: Error resilient transmission for compressed sensing of color images with multiple description coding. In: Proceedings of International Conference on Robot Vision and Signal Processing, pp. 63–66 (2015)

10. Wang, Z., Bovik, A.C., et al.: Image quality assessment: from error visibility to structural similairty. IEEE Trans. Image Process. **13**, 600–612 (2004)

Camellia Key Expansion Algorithm
Based on Chaos

Chuanfu Wang[(✉)] and Qun Ding

Institute of Electrical Engineering, Heilongjiang University,
Xuefu road no. 74, Harbin 150080, Heilongjiang, China
{chuanfuwang, qunding}@aliyun.com

Abstract. The shortage of original Camellia key generation and expansion strategy was researched. Based on these researches, the new Camellia key expansion algorithms are proposed. In this algorithms, Logistic mapping is used to reduce the dependence between subkeys. Experimental result shows that the security and robustness of the AES subkeys have been strengthened.

Keywords: FPGA · Camellia · Key scheme

1 Introduction

Over the past decade, the rapid expansion of digital communication technologies has been simply astounding. With the advent of the information age, big data attracts more and more attention. During the process of big data, data security transmission guarantees the reliability of information. So, cryptography play an important role in the digital communication system. In recent years, the requirement of information security is enhanced. So, a lot of encryptions are presented, such as DES, AES, Camellia and so on. In cryptography, Camellia is a Block cipher that is highly praised by many organizations, including the European Union's NESSIE project (as a selection algorithm) and the Japanese CRYPTREC project (as a recommendation algorithm). Camellia algorithm was designed by the NTT and Mitsubishi Electronics Company. It was first published in the 2000 SAC conference, and then became the European encryption standard. Camellia and AES have the same level of security strength and computing. Camellia is widely used in many European security systems.

Chaos is a new emerging discipline in recent decades, which has a lot of characteristics, such as initial value sensitivity, ergodic, deterministic and so on. These characteristics are equal to the 'diffusion' and 'confusion', which were proposed in the cryptography by Shannon. Chaos mapping have very complex nonlinear characteristic and the output sequences have strong randomness, which are often used in stream cipher as pseudo-random sequence.

Block cipher encrypts data through several iterations. In each round, subkeys are required to participate in the operation. Based on the idea of cryptography attack, subkeys will be found out through the comparison of the plaintext pair and the ciphertext pair. Initial key will be deciphered finally. The most commonly used cryptographic attacks include differential attack and linear attack. In order to avoid

© Springer International Publishing AG 2017
J. Pan et al. (eds.), *Intelligent Data Analysis and Applications*, Advances in Intelligent Systems and Computing 535, DOI 10.1007/978-3-319-48499-0_26

being subjected to differential and linear attack, subkeys should be better spread in the iterative algorithm in each round. Therefore, a new key expansion algorithm is proposed to enhance the randomness of subkeys in each round via combining the Camellia algorithm and the chaos mapping. So, the subkeys can be better spread in the iterations.

2 Camellia

Camellia algorithm is a Block cipher. The length of Camellia algorithm plaintext packet are 128 bits. Secret key have three choice to choose, including the length of 128, 192 or 256 bits. According to the different length of the secret key, the number of rounds were 18, 24 and 24. One of the unique features of the Camellia is the design of the diffusion layer, which just consist only of bit XOR operation. Feistel structure is adopted in Camellia algorithm, and the round function F is composed of two modules, which are confusion and diffusion.

2.1 Confusion

Confusion module always be completed by S-box, which is the only nonlinear component in many Block ciphers. Therefore, the strength of the S-box determines the strength of the whole cryptosystem, and its efficiency determines the efficiency of the whole cryptosystem. Camellia has four different S-box. Those S-boxes are completed by the inverse function over the finite field and the affine transformation. The S-box2, S-box3 and S-box4 are derived from the shift operation of the S-box1.The whole S-boxes algebraic expressions are as follows

$$s - box1 : F_2^8 \rightarrow F_2^8 \qquad x_{(8)} \rightarrow h(g(f(x_{(8)} \oplus 0xc5)) \oplus 0x6e \tag{1}$$

$$s - box2 : F_2^8 \rightarrow F_2^8 \qquad x_{(8)} \rightarrow s - box1(x_{(8)}) <<<1 \tag{2}$$

$$s - box3 : F_2^8 \rightarrow F_2^8 \qquad x_{(8)} \rightarrow s - box1(x_{(8)}) >>>1 \tag{3}$$

$$s - box4 : F_2^8 \rightarrow F_2^8 \qquad x_{(8)} \rightarrow s - box1(x_{(8)} <<<1) \tag{4}$$

In S-box1, function F is given by $f : B \rightarrow B$

$$
\begin{bmatrix} B_1 \\ B_2 \\ B_3 \\ B_4 \\ B_5 \\ B_6 \\ B_7 \\ B_8 \end{bmatrix}
=
\begin{bmatrix}
0 & 0 & 0 & 0 & 0 & 1 & 0 & 0 \\
1 & 0 & 0 & 0 & 0 & 0 & 1 & 0 \\
0 & 0 & 1 & 0 & 0 & 0 & 0 & 1 \\
0 & 0 & 1 & 0 & 0 & 0 & 0 & 1 \\
0 & 0 & 0 & 1 & 0 & 0 & 1 & 0 \\
0 & 1 & 0 & 0 & 1 & 0 & 0 & 0 \\
1 & 0 & 0 & 0 & 0 & 0 & 0 & 1 \\
0 & 0 & 0 & 1 & 0 & 1 & 0 & 0
\end{bmatrix}
\begin{bmatrix} A_1 \\ A_2 \\ A_3 \\ A_4 \\ A_5 \\ A_6 \\ A_7 \\ A_8 \end{bmatrix}
\tag{5}
$$

Function h is given by $h : B \rightarrow B$

$$
\begin{bmatrix} B_1 \\ B_2 \\ B_3 \\ B_4 \\ B_5 \\ B_6 \\ B_7 \\ B_8 \end{bmatrix} = \begin{bmatrix} 0 & 1 & 0 & 0 & 1 & 1 & 0 & 0 \\ 0 & 1 & 0 & 0 & 0 & 0 & 1 & 0 \\ 0 & 0 & 0 & 1 & 0 & 0 & 1 & 0 \\ 0 & 1 & 0 & 0 & 0 & 0 & 0 & 1 \\ 0 & 0 & 1 & 0 & 0 & 0 & 1 & 0 \\ 1 & 0 & 0 & 0 & 0 & 0 & 0 & 1 \\ 1 & 0 & 0 & 0 & 1 & 0 & 0 & 0 \\ 0 & 0 & 1 & 0 & 0 & 1 & 0 & 0 \end{bmatrix} \begin{bmatrix} A_1 \\ A_2 \\ A_3 \\ A_4 \\ A_5 \\ A_6 \\ A_7 \\ A_8 \end{bmatrix} \tag{6}
$$

Function g is given by $g : B \rightarrow B$ $(B_1 B_2 B_3 B_4 B_5 B_6 B_7 B_8) = g(A_1 A_2 A_3 A_4 A_5 A_6 A_7 A_8)$

Function g is an operation over the finite field. It's so difficult to compute the formula of Function g.

2.2 Diffusion

The role of the diffusion layer is to scramble and mix the output of these S-boxes. So, the output of the 8 bits are as much as uncorrelated with other S-box input which are related. Diffusion layer is usually achieved by a nonlinear transformation function of L. Since the operation of confusion is composed of 4 S-box parallel, L can be regard as a transformation, which algebraic expressions can be displayed as $(F_2^8)^4 \rightarrow (F_2^8)^4$. So, the diffusion layer of the Camellia algorithm matrix is shown in (7) in detail.

$$
A = \begin{bmatrix} 0 & 1 & 1 & 1 & 1 & 0 & 0 & 1 \\ 1 & 0 & 1 & 1 & 1 & 1 & 0 & 0 \\ 1 & 1 & 0 & 1 & 0 & 1 & 1 & 0 \\ 1 & 1 & 1 & 0 & 0 & 0 & 1 & 1 \\ 0 & 1 & 1 & 1 & 1 & 1 & 1 & 0 \\ 1 & 0 & 1 & 1 & 0 & 1 & 1 & 1 \\ 1 & 1 & 0 & 1 & 1 & 0 & 1 & 1 \\ 1 & 1 & 1 & 0 & 1 & 1 & 0 & 1 \end{bmatrix} \tag{7}
$$

2.3 Function FL and FL^{-1}

Function FL and FL^{-1} algebraic expressions are as follows

$$
FL : F_2^{64} \times F_2^{64} \rightarrow F_2^{64} \quad (X_{L(32)} \parallel X_{R(32)}, kl_{L(32)} \parallel kl_{R(32)}) \rightarrow Y_{L(32)} \parallel Y_{R(32)} \tag{8}
$$

$$
\text{Where } Y_{R(32)} = ((X_{L(32)} \cap kl_{L(32)}) <<<1) \oplus X_{R(32)} \tag{9}
$$

$$Y_{L(32)} = (Y_{R(32)} \cup kl_{R(32)}) \oplus X_{L(32)} \tag{10}$$

$$FL^{-1} : F_2^{64} \times F_2^{64} \to F_2^{64} \quad (Y_{L(32)} \parallel Y_{R(32)}, kl_{L(32)} \parallel kl_{R(32)}) \to X_{L(32)} \parallel X_{R(32)} \tag{11}$$

$$\text{Where } X_{L(32)} = (Y_{R(32)} \cap kl_{R(32)}) \oplus Y_{L(32)}$$
$$X_{R(32)} = ((X_{L(32)} \cap kl_{L(32)}) <<<1) \oplus Y_{R(32)} \tag{12}$$

2.4   Round Funtion F

Round Function F is the basic operation in Camellia algorithm. The Round Function F is shown in Fig. 1.

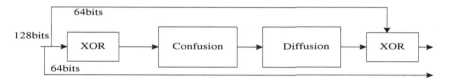

Fig. 1.   Round Function F

2.5   Encryption Algorithm

The encryption algorithm is shown in Fig. 2. Camellia algorithm include 18 rounds of the Round Function F.

Subkeys is generated by the key expansion algorithm, and each round subkeys is different. Each round subkeys is obtained by cyclic shifting different bits through the KL. KL in the Fig. 2 means the output of key expansion algorithm. Key expansion algorithm is just like encryption algorithm but have less round iterations. Key expansion algorithm is shown in Fig. 3.

$M1 = 0xA09E667F3BCC908B, \quad M2 = 0XB67AE8584CAA73B2,$
$M3 = 0XC6EF372FE94F82BE, \quad M4 = 0X54FF53A5F1D36F1C$

3   Chaos

Chaos is a kind of determined dynamics system, which is not predictable. Just a small difference of the initial value changed, the output of the Logistic Mapping will be very complex and unpredictable owing to the sensitivity to the initial value. Logistic mapping is used here for an easy hardware implementation. Logistic mapping is one of the easiest chaotic mapping, because it is very easy to be achieved by hardware. Logistic mapping formula is as follows:

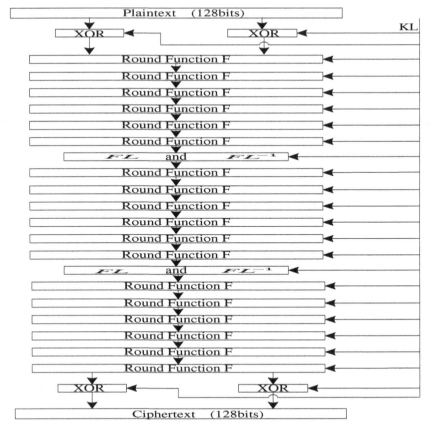

Fig. 2. Encryption algorithm

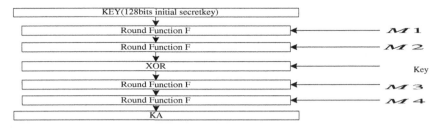

Fig. 3. Key expansion algorithm

$$X_{n+1} = \mu X_n(1 - X_n) \quad \mu \in (0, 4], X_n \in (0, 1) \qquad (14)$$

μ is a parameter of the logistic mapping, which controls the chaotic behavior. X_n is the initial value of the Logistic mapping, and affects the orbit of chaos. Chaotic orbit is sensitive to the initial value extremely. Even a small change in initial value can make a

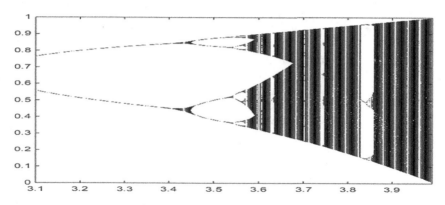

Fig. 4. The bifurcation diagram of the Logistic mapping

great difference in chaotic orbit. The bifurcation diagram of the Logistic mapping is shown in Fig. 4

As shown in Fig. 4, when the value of μ is located in $(3.5699, 4]$, the chaotic characteristic is obvious in Logistic mapping. When the value of μ is chosen to be 4, the orbit of Logistic mapping has filled the entire range. At that time, the best characteristics behavior in the Logistic mapping is presented. The complex dynamic characteristics of the Logistic mapping determine the output must be a pseudo-random sequence. So, Camellia key expansion algorithm can be replaced by the Logistic mapping, because it has better pseudorandom characteristics.

4   Improved Camellia Algorithm Based on Logistic Mapping

As we discussed in the key expansion, KL is the output of the key expansion algorithm. Each round of subkeys is obtained only by cyclic shifting KL differently. KL is just calculated via four rounds of the round function F. Hence, traditional public Camellia encryption algorithm has safety problem. In order to make up for the shortage of Camellia algorithm in the aspect of key expansion, the pseudo-random sequences generated by the Logistic mapping can improve the subkeys obtained by cyclic shifting from KL. The pseudo-random sequences generated by the chaotic system are with great random properties, and often applied in the design of stream cipher. The improved Camellia is shown in Fig. 5.

Logistic mapping in Fig. 4 is an expression in the real number field, which needs to be quantified in order to be applied to the digital system. Since the hardware system can't directly represent the decimal, and software implementation of the logistic mapping also need to be quantified as two valued sequence. Only the two valued sequence can be involved in arithmetic operation. So using the fixed-point number to replace the float-point number, the output pseudo-random sequences of the Logistic mapping were obtained by FPGA. Hardware circuit diagram of the Logistic mapping is shown in Fig. 6

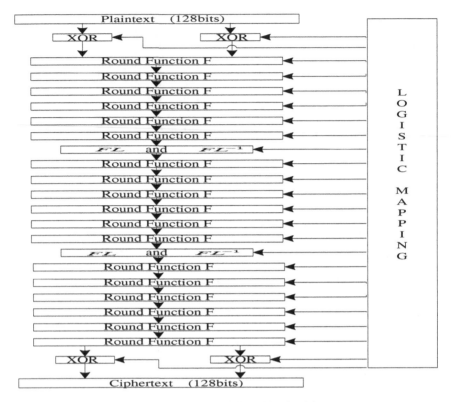

Fig. 5. The improved Camellia algorithm

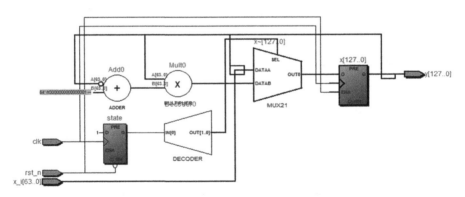

Fig. 6. Hardware circuit diagram of the Logistic mapping

As seen in Fig. 6, the numbers in the real number field is represented by 64 bits binary numbers. The quantized Logistic mapping's output are represented by the 128 bits binary numbers, which is not like software quantification obtained bit by bit. Statistics show that the output of the 128 bits numbers in each iteration rounds are independent, and with good random characteristics. Therefore, the pseudo-random

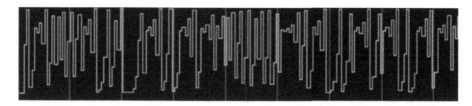

Fig. 7. Pseudo-random sequences of the Logistic mapping

sequence implemented by the hardware is 128 times than the software. The output pseudo-random sequences of the Logistic mapping was gotten by the simulation software Modelsim. The output pseudo-random sequences of the Logistic mapping is shown as Fig. 7.

In the Fig. 7, each pulse is represented by the 128 bits, and the height of the pulse is represented by a 128 bits binary value. The fluctuation of the pulse height shows the random characteristic of the output of the Logistic mapping. As shown in Fig. 8, Relative to the subkeys achieved by cyclic shifting KL by different bits, the pseudo-random sequences generated by the logistic mapping's random characteristic is better.

5   Conclusion

For solving the safety problems of existed Camellia algorithm, an improved Camellia algorithm based on Logistic mapping is proposed in this paper. The subkeys generated by the Logistic mapping is completed by the FPGA and simulated by the Modelsim. The output are proved to be a good random characteristics

References

1. Aoki, K., Ichikawa, T., Kanda, M., Matsui, M., Moriai, S., Nakajima, J., Tokita, T.: Specification of Camellia – A 128-bit Block Cipher. Nippon Telegraphy and Telephone Corporation, Mitsubishi Electric Corporation, July 2000
2. Aoki, K., Ichikawa, T., Kanda, M., Matsui, M., Moriai, S., Nakajima, J., Tokita, T.: Camellia: a 128-bit block cipher suitable for multiple platforms - design and analysis. In: Stinson, D.R., Tavares, S. (eds.) SAC 2000. LNCS, vol. 2012, pp. 39–56. Springer, Heidelberg (2001)
3. Zhou, Y., Wu, W., Xu, N., Feng, D.: Differential fault attack on camellia. Chin. J. Electron. **18**(1), 13–19 (2009)
4. Kocarev, L., Jakimoski, G., Stojanovski, T., Parlitz, U.: From chaotic maps to encryption schemes. In: Proceedings of IEEE International Symposium on Circuits and Systems 1998, vol. 4, pp. 514–517. IEEE (1998)
5. Zhou, H., Ling, X.T.: Generating chaotic secure sequences with desired statistical properties and high security. Int. J. Bifurcation Chaos **7**(1), 205–213 (1997)

6. Li, S.J.: Analyses and new designs of digital chaotic ciphers. Ph.D. thesis, School of Electronics and Information Engineering. Xi'an Jiaotong University, Xi'an (2003). http://www.hooklee.com/pub.html

7. Li, C.Y., Chen, J.S., Chang, T.Y.: A chaos-based pseudo random number generator using timing-based reseeding method. In: IEEE Proceedings of ISCAS, pp. 21–24 (2006)

8. Hatano, Y., Sekine, H., Kaneko, T.: Higher order differential attack of camellia (II). In: Nyberg, K., Heys, H.M. (eds.) SAC 2002. LNCS, vol. 2595, pp. 39–56. Springer, Heidelberg (2003)

9. Shirai, T., Kanamaru, S., Abe, G.: Improved upper bounds of differential and linear characteristic probability for camellia. In: Daemen, J., Rijmen, V. (eds.) FSE 2002. LNCS, vol. 2365, pp. 128–142. Springer, Heidelberg (2002)

10. Zhang, L., Zhang, Y.: Research on lorenz chaotic stream cipher. In: IEEE International Workshop VLSI Design & Video Tech, pp. 431–434 (2005)

11. Lu, J., Kim, J.-S., Keller, N., Dunkelman, O.: Improving the efficiency of impossible differential cryptanalysis of reduced camellia and MISTY1. In: Malkin, T. (ed.) CT-RSA 2008. LNCS, vol. 4964, pp. 370–386. Springer, Heidelberg (2008)

12. Addabbo, T., Alioto, M., Fort, A., Pasini, A., Rocchi, S., Vignoli, V.: A class of maximum-period nonlinear congruential generators derived from the Rényi chaotic map. IEEE Trans. Circuits Syst. I **54**(4), 816–828 (2007)

13. Li, G., Wang, Y., Ma, H.: Audio packet encryption of flow control base on stop-and-wait protocol. J. Inf. Hiding Multimed. Signal Process. **7**(1), 127–134 (2016)

14. Zhen, J., Zhifang, W., Zhao, B., Tang, L.: A novel super-resolution direction finding method for wideband signals based on multiple small aperture subarrays. J. Inf. Hiding Multimed. Signal Process. **7**(3), 510–520 (2016)

The Key Exchange Algorithm in Network Encryption Machine

Minghao Li[1] and Qun Ding[2(✉)]

[1] The Secure Communication research room, Key Lab Electronic Engineering,
College of Heilongjiang Province, Harbin, China
[2] A professor and doctoral tutor of College of Electronic Engineering,
Heilongjiang University, Harbin, People's Republic of China
2151349@s.hlju.edu.cn

Abstract. This paper mainly presents a key exchange realization method about network encryption machine. The way is based on Diffie-Hellman Key Exchange and Authenticated Technology to make all network encryption machine increasingly security.

Keywords: Diffie-Hellman Key Exchange · Network · Certification

1 Introduction

In 21st century, information transmission has become an important part of our life, moreover, great attention is paid to the information security. The primary reason is that when the information leaks out and then the serious leakage would have damaging impart on personal property and country safety. Therefore, the study of security of information transmission has become increasingly important.

One of important aspects to assure the information security is cryptographic transmission. The way can protect our transmission content and increase risks of information leakage. The most important of cryptographic transmission is secret-key security, and the secret-key is the "key" of encryption technology and decryption technology. Traditional cryptographic transmission system such as Fig. 1 introduces security of cryptographic transmission.

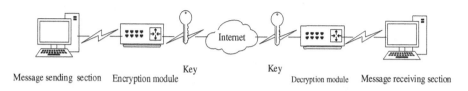

Message sending section   Encryption module   Key   Key   Decryption module   Message receiving section

Fig. 1. The schematic of the transmission system by using encrypted with a key

Although there are diverse programs of cryptographic transmission nowadays, the fundamental principle is divided two branches—channel encryption and data encryption. On one hand, channel encryption is encryption processing of physical layer signal.

© Springer International Publishing AG 2017
J. Pan et al. (eds.), *Intelligent Data Analysis and Applications*, Advances in Intelligent Systems and Computing 535, DOI 10.1007/978-3-319-48499-0_27

It would provide processing methods of channel encryption and Bit-Torrent, but decoding method is provided by user as secrecy basis. However, there are some problems about the way, for example, all information will leak out based on the prerequisite—unsafe channel. On the other hand, data encryption is the transmission of real-time data transmission to go to encrypt processed, so that even in the absence of protective measures can also achieve safe and efficient channels of information transmission. Key encryption described above is one commonly used method of data encryption transmission. Therefore, this paper introduces a data encryption as research background (Fig. 2).

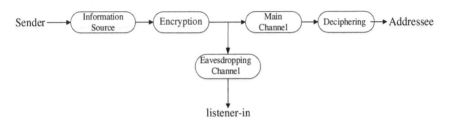

Fig. 2. The diagram of channel encrypted transmission system

2   The Key Exchange

2.1   General Introduction

Data encryption is one way transmission of information encryption method encryption method proposed after relatively long been used in general, including symmetric encryption, public-key block cipher encryption and password encryption. Briefly the core of symmetric encryption method is to encrypt and decrypt using the same key; block cipher encryption method is mainly used an iterative approach to data encryption mode, shown as Fig. 3; public key cryptography is encryption has two keys: a public key and private key that is the method of combining encrypted with the public and private key to decrypt.

Three common ways a mixed blessing, first introduced symmetric encryption, which operating speed is relatively high, but the key is a single, once leaked information will leak, security is gone; and then look at the block cipher encryption mode plaintext good scalability, and does not require a key synchronization, more suitable as an encryption standard, but the method should not be overlooked drawback is that encryption is slow

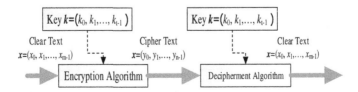

Fig. 3. Schematic diagram of the structure of packet encryption

and prone to error diffusion and dissemination; encryption called public-key cryptography is the most secure encryption method one, two key fully guarantee the security of the encryption tools, making information security has been further guaranteed; but far more virtues than defects, it is also the least efficient encryption method, resulting in greater cost of the way, it will not be universally applicable.

So, we decided to only the key from the security key exchange mode, take information block cipher encryption method for encryption and decryption. So it can make good use of the advantages and disadvantages of several data encryption transmission was efficient and safe transport.

2.2 Introduction of Key Exchange

Key exchange is an important step in this whole machine network encryption link encryption algorithm. This concept was first proposed by Ralph C. Merkle, US cryptographers Bailey Whitfield Diffie and Martin Edward Hellman first published in 1976, the implementation method of key exchange, which is mainly in the transport layer to enhance information security. As more practical use of the theory, a lot of information security systems are gradually update their key security system. Currently the key exchange method practicability relatively strong mainly non-symmetric encryption algorithm RSA, Diffie-Hellman algorithm (DH algorithm for short) and a digital signature method.

We use the Diffie-Hellman algorithm uses on our network encryption link machine, the main principle is the number of large prime after a random number the user input or automatically generated the corresponding operation was used as a transmission key used.

2.3 The Principle of Diffie-Hellman Algorithm

In 1976, American cryptographers Bailey Whitfield Diffie and Martin Edward Hellman put forward to key exchange method is the D-H algorithm, and because of Ralph C. Merkle key exchange is the author of the concept, which is also known as Diffie-Hellman - Merkle key exchange. The paper referred to this method as the D-H algorithm.

D-H algorithm mainly through the difficulty of discrete number of large prime number algorithm uses a large number of bits on math, it is difficult to be fast cracks, can guarantee its own security key. Theorem mainly through Miller - Rabin prime principle of choice, in addition to re-screen after a composite number and Carmichael number of large prime numbers would come to meet the conditions, and then repeated by the mold (square calculation) method can put complicated remainder recursive computation of large integer mode m and n are transformed into a simple decomposition operation (large numbers into a binary representation of the decom-position step to calculate large prime numbers), the calculation result is that we want the key, the algorithm is a schematic diagram shown in Fig. 4.

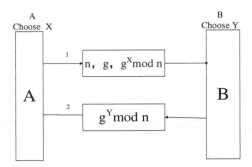

Fig. 4. The schematic of Diffie-Hellman algorithm

2.4 The Implementation Steps of Diffie-Hellman Algorithm

This section describes the implementation steps DH algorithm steps shown in the schematic in Fig. 5. The following narrative will be divided into five steps to describe:

The first step: Sender A and addressee B must also know that the two large prime numbers: n and g, and $(n-1)/2$ and $(g-1)/2$ is a prime number. These figures are also public, so anyone can choose two among the n and g and openly tell another person.

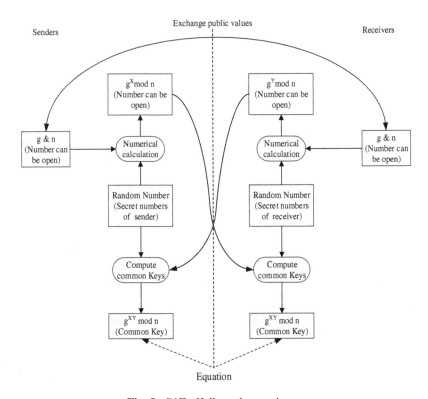

Fig. 5. Diffie-Hellman key exchange.

Step Two: A sender selected a large number X (for example: 512 bits), and the secrecy it up the same way, the addressee B selected a large number of secret Y.

The third step: A sender and addressee to start with B key exchange, he was first sent to the other party including the information a B (n, g, g^X mod n).

Step four: addressee B sent a message to sender A containing the g^Xmod n to do answer.

Step Five: A sender send their digital to addressee B and solve X-th power demand calculation as in (1), B performs a similar operation as in (2). Thus, the sender and the addressee A to B share a secret key g^{XY}mod n

$$(g^Y \bmod n)^X = g^{XY} \bmod n \tag{1}$$

$$(g^X \bmod n)^Y = g^{XY} \bmod n \tag{2}$$

3 Security Certificate

While key exchange can be transferred only public figure in the channel rather than the key itself, so that a third party, even if hackers intercept digital transmission does not help, you cannot get the key. But with the widespread use of key exchange, a new expose the problem, that is, the attacker disguise, pretending sender or receiver that is sending a confusing fake number, so that the other figures with this fake a large prime number as a key, Although the prime meet Criterion D-H algorithm, but the attacker has been wholly obtained, then use this key to encrypt the information is very dangerous. So we need to take measures to determine the authentication data sender or receiver received exchange key is not to send each other.

3.1 Overview

Certification is a very common way of confirmation in our life, information transmission technology also requires authentication, such as in our key exchange is required to sender or receiver coming digital certification: the other side is coming to normal use (the digital power of the corresponding request, obtain key), or we need to remove the attacker, to avoid falling into the key hands of hackers.

Commonly used authentication methods are one-way hash functions, message authentication codes, digital signature and certificate authentication, these methods have their own field of use. One-way hash function is mainly to verify the integrity of the file, the file is force majeure factors prevent damage to the user without being aware, affect the use; message authentication code is an important tool to determine whether the message is properly conveyed; digital signature is OK message sender who is the main tool, that is, who wrote the message? Certificate is for the public key plus a digital signature to ensure public safety. DH key exchange authentication schematically shown in Fig. 6.

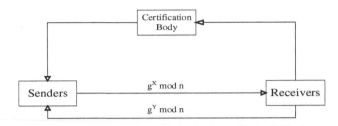

Fig. 6. The certification of Diffie-Hellman Key exchange

3.2 The Certificate Authentication

Based on third-party the certificate authentication, can effectively prevent intruders attack. Control Fig. 6 D-H key exchange authentication shown, we can see concrete steps to certificate authentication.

The first is the information reception aspect generates a key pair to be generated by the certification body on behalf; Further information about the recipient's public key registered in their own certification bodies, and certification bodies will confirm the reliability of the public; and certification bodies using the private key of the digital signature is applied to the information coming from the recipient and generate public security certificate; at this regard will be sent information with the public key certification authority digital signature; the public key to verify the last message sender uses the certification body digital signature, the public key to verify the legitimacy of the receiving party, thus ensuring both send and receive information on the safe and smooth data transfer.

4   Software Implementation

4.1   Program Introduction

D-H algorithm code in the process of running first need to randomly selected large prime numbers p and $G_F(p)$ of the primitive root g, and p and g as public information; when the user wants to send a message and receive users to establish secure communication needs exchange and sharing of key K_{AB}, the sender selected at random integer X_A $(1 < X_A < p-1)$, and calculates the $Y_A = g^{XA}$ mod p, then passed to the recipient through the common channel, the X_A confidential; recipients also randomly selected 1 integer $X_B(1 \leq X_a \leq 1)$, and calculate the $Y = g^{XB}$ mod p, also pass through the common channel sender, the X_B secrecy; the last sender and recipient each calculation $K_{AB} = Y_A X_B$ mod p and $p = Y_B X_A$ mod p, Thus the sender, recipient successfully exchanged secret key K_{AB}. Program flow shown in Fig. 7.

4.2   Efficiency Comparison

The program D-H key exchange algorithm with our newly designed implant network encryption machine. Then connect to the Internet, run through the program so as to

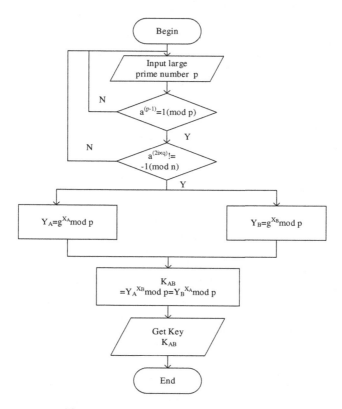

Fig. 7. The program flow chart of algorithm

realize transmission of data encryption, encrypted configuration parameters as shown in Fig. 8. And then measuring the time of this method of encryption and decryption. To compare the efficiency of non-key exchange algorithm when transferring data, namely encryption and decryption time used, we have to generate 512 Key Case contrast, comparison of Table 1, you can observe the feasibility of this method.

Table 1. Use 512-bit key to encrypt information with time

Algorithm number	Non key exchange algorithm		After adding the key exchange algorithm	
	Encrypt	Decrypt	Encrypt	Decrypt
1	1.73	1.82	2.15	2.23
2	1.80	1.77	2.08	2.21
3	1.82	1.74	2.11	2.17
Average	1.79	1.79	2.11	2.21

a) Encryption terminal   b) Decryption terminal

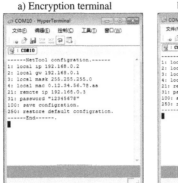

Fig. 8. Physical map of the network encryption module

References

1. Weng, S., Pan, J.-S., Deng, J.: Invariability of remainder based reversible watermarking. J. Netw. Intell. **1**(1), 16–22 (2016)
2. Diffie, W., Hellman, M.E.: New directions in cryptography. IEEE Trans. Inf. Theor. **22**, 638–654 (1976)
3. Rojas, D.A., Ramos, O.L., Saby, J.E.: Recognition of spanish vowels through imagined speech by using spectral analysis and SVM. J. Inf. Hiding Multimed. Sig. Process. **7**(4), 889–897 (2016)
4. Guttman, J.D.: State and progress in strand spaces: proving fair exchange. J. Autom. Reason. **48**(2), 159–195 (2012)
5. Garg, S., Gentry, C., Halevi, S., Raykova, M., Sahai, A., Waters, B.: Candidate in distinguish ability obfuscation and functional encryption for all circuits. In: Proceedings of FOCS (2013)
6. Zomaya, A.Y., Lee, Y.C.: Energy conscious scheduling for distributed computing systems under different operating conditions. IEEE Trans. Parallel Distrib. Syst. 22, 1374–1381 (2011)
7. Groce, A., Katz, J.: A new framework for efficient password based authenticated key exchange. In: Proceedings of the 17th ACM Conference on Computer and Communications Security (CCS 2010), pp. 516–525, Chicago, USA (2010)
8. Sun, Y., Zhu, H.F., Feng, X.: An enhanced dragonfly key exchange protocol using chaotic maps. J. Inf. Hiding Multimed. Sig. Process. **7**(2), 376–385 (2016)
9. Zhao, J., Gu, D.: Provably secure authenticated key exchange protocol under the CDH assumption. J. Syst. Softw. **83**, 2297–2304 (2010)
10. Xin, Y., Shun, W.Z., Gong, C.R.: Application of group key agreement based on authenticated Diffie-Hellman for bluetooth Piconet. In: WASE International Conference on Information Engineering, ICIE 2009, pp. 125–128 (2009)
11. Hao, F., Clarke, D.: Security analysis of a multi-factor authenticated key exchange protocol. In: Bao, F., Samarati, P., Zhou, J. (eds.) ACNS 2012. LNCS, vol. 7341, pp. 1–11. Springer, Heidelberg (2012)
12. Cortier, V., Dougherty, D.J., Guttmann, J.D., Kremer, S.: The real-or-random property for lightweight Diffie-Hellman protocols. Meetings, January 2014

Automatic Detection Method for Dynamic Topology Structure of Urban Traffic Network

Xianghai Ge[(⊠)], Xinhua Jiang, Fumin Zou, and Lvchao Liao

The Key Laboratory for Automotive Electronics and Electric Drive of Fujian Province, Fujian University of Technology, Fuzhou 350018, China
{haizige,xhjiang,fmzou,achao}@fjut.edu.cn

Abstract. In order to further explore and discover the inner structure features of complex urban traffic network, based on urban road network geography model and the mass of dynamic traffic trajectory data resources, proposed a kind of dynamic topology structure automatic discovery method of urban traffic network. Firstly based on urban geographical model, and construct the graph model of urban road network, that is taking road intersection as a node, and regard connected path between nodes as links, establish graph model of node link to represent the topological structure of urban road network, further based on the running road speed information of collected road of floating car data collection added weight for the link, and build the weighted directed graph topology of urban traffic network. Taking Nanning city as an example, the experimental results show that this method can find its intrinsic topology characteristics and realize the dynamic updating of urban road network topology.

Keywords: Traffic road network · Road network structure · Topology · Weighted directed graph

1 Introduction

The scientific planning and effective operation of traffic network for the stable development of the city plays a vital role, and the core and the key to realize scientific planning and management of transportation network is in-depth understanding of urban traffic network and dynamic structure characteristics of traffic flow [1]. Therefore, it is urgent to find an effective method to explore and find the dynamic topology structure of urban traffic network, which provides an important foundation for the analysis of the traffic network topology and the optimization of the road network.

Urban road network is a complex dynamic system, how to effectively discover the features of intrinsic topology is one of the key problems in the field of transportation planning and management [2–5]. At present, a series of researches have been carried out on this issue at home and abroad, and the important research results have been obtained. Among them, the literature [6] is based on the mass of the traffic trace data to extract the important intersection of urban traffic network as the urban node, but not build the whole road network topology. The literature [7] proposed a discovery method based on the mass trajectory data of traffic network topology, but completely stay away from the support of GIS map data, processing precision is limited. the literature [8]

© Springer International Publishing AG 2017
J. Pan et al. (eds.), *Intelligent Data Analysis and Applications*, Advances in Intelligent Systems and Computing 535, DOI 10.1007/978-3-319-48499-0_28

presents a graph model of the dynamics of traffic network, its main function is to achieve the traffic network topology quickly storing and updating, literature [9] is based on traffic network topology analyzed road traffic vulnerability, the literature [10] present an association probability characteristics between roads based on the link mechanism (PageRank) similar to Google Docs. Although existing research has made a series of achievements for urban traffic network topology discovery and application, but in the face of the complex urban road network, there is a lack of effective mechanism to quickly extract the topology information, and updating the dynamic online.

In order to discover the complicated urban road network topology information effectively, based on urban road network geography model and the mass of dynamic traffic trajectory data resources, this paper presents a kind of automatic discovery method of dynamic topology structure of urban traffic network. Firstly based on urban geographical model, and construct the graph model of urban road network, that is taking road intersection as a node, and regard connected path between nodes as links, establish graph model of node link to represent the topological structure of urban road network, and further based on the road network of massive traffic track data and accessibility and travel time attribute of dynamic mining link, and then construct the urban traffic network weighted directed graph topology. Taking Nanning city as an example, the experimental results show that this method can find its intrinsic topology characteristics and realize the dynamic updating of urban road network topology.

2 Dynamic Topology Modeling of Urban Road Network

The urban road network is complex, and there are a lot of complex relationships among roads, such as intersection, overlap and spatial height deviation. And the flow characteristics of the existing research road mainly based on road analysis, often only observed local road traffic information, and are difficult to find whole structure characteristics of the road traffic, especially because of adding the viaduct to make the systematic modeling based on the road becomes more difficult. Thus, this article takes the actual intersection as a node between roads, and regard the directed and reachable roads at every road node as links, and construct the urban road network topology model based on nodes and links (Fig. 1).

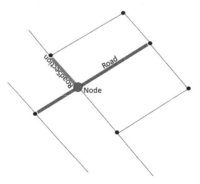

Fig. 1. Schematic diagram of urban network topology

2.1 Dynamic Topology Modeling of Urban Road Network

According to the thoughts of graph theory, urban road network topology are expressed a directed weighted graph of "node – link" which is composed of three elements of the grounds of node, link, weight. That is, complex urban traffic network is abstracted a containing the weights of the graph model, the specific dynamic network topology models are:

$$T = (N, R, W)$$

where,

T represents a structure model of urban road network topology;

$N = \{n_i | i = 1, 2, 3, \ldots, n\}$ represents a collection of nodes in a city road network;

$R = \{r < n_i, n_j > | i, j \in N\}$ represents a collection of road segments in the urban road network, $r < n_i, n_j >$ refer to sections (groups) from node i to node j;

$W = \{w < r > \}$ represents a set of weights in the middle sections of the urban road network, $w < r >$ refers to the weight value of the road segment.

In this paper, the road intersection is used as the research entrance of the road network topology structure, and the intersection is abstracted as the node, so interactions of road segment are follows:

$$C = (R_s, L)$$

$C = \{c_i | i = 1, 2, 3, \ldots, n\}$ is a collection of intersections in urban road networks;

R_s is cross section of road, and L is position coordinates for intersections.

At the same time, in order to further express inter node connectivity here introduced a dynamic link weight factor, make travel time of link as the weight, namely obtained by the calculation of the link length and floating car speed of road, and also can be calculated combined with the historical data. But generally weight is dynamic, to achieve network topological properties matching with practical application scene.

2.2 Model and Representation of Road Network Topology

Based on modeling method of dynamic network topology, the urban complex network is represented as a node-link weighted graph and by the connection structure of each node to analyze the intrinsic topological characteristics of urban traffic network. At the same time, the urban transportation network is abstracted as a weighted directed graph topology, topology problem is transformed into the directed weighted graph problems. The weight information of the directed weighted graph has dynamic and specific application scenarios preferences, in order to facilitate the expression and renewal of topological weight information of the urban road network. In this paper, represent weight value by the adjacency matrix.

Adjacency matrix is a matrix that represents the adjacency relationship between nodes in the graph. If there are n nodes in the graph, the n*n matrix can be constructed to represent the connection relationship between the N nodes. In general, for the

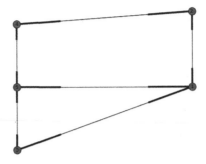

Fig. 2. Road network topology example

adjacency matrix, if the node i and j connected, then the value of the representation matrix (I, J) of the item is 1, otherwise is 0. That is:

$$A(i,j)=\begin{cases} 1, i \ to \ j \ Connected \\ \infty, \qquad otherwise \end{cases}$$

The adjacency matrix of Fig. 2 can be expressed as

i,j	1	2	3	4	5
1	0	1	1	0	1
2	1	0	0	1	0
3	1	0	0	0	1
4	0	1	0	0	1
5	1	0	1	1	0

For the adjacency matrix with link weights, the 1 of the adjacency matrix is replaced by the link weight, replaced the unreachable 0 value as infinite value:

$$A(i,j)=\begin{cases} w_{ij}, i \ to \ j \ Connected \\ \infty, \qquad otherwise \end{cases}$$

The weighted adjacency matrix of Fig. 2 can be expressed as:

i,j	1	2	3	4	5
1	∞	3	8	v	4
2	3	∞	∞	4	∞
3	6	∞	∞	∞	6
4	∞	5	∞	∞	4
5	4	∞	5	4	∞

3   Algorithm Implementation

3.1   Pretreatment of Urban Road Network Node

The dynamic topology of urban road network consists of three elements: nodes, links and their weights. Among them, the extraction of network node is the key link. Mainly selected each road layer, and analysis the intersection characteristic that the actual achieved between roads, and then extract the reachable intersection node of the road as topology, while unreachable intersection point, such as Viaduct with elevation plane intersection points do not regard as a node.

Because of the quality of each map data and specifications are uneven, in order to ensure the algorithm's versatility. Firstly, we extract starting and ending point of all the sections of the road network layer, provide pretreatment data for further extraction of urban road network node.

Specific algorithm is:

Step 1:   Obtain traffic road data of the various layers of the map;

Step 2:   Access to the start point Ps of road segment and the end point Pe and the length of the section, angle and other information;

Step 3:   Calculate the distance Dci between the point Ps and point Pe and all nodes in the connection point list C;

Step 4:   If the Dci is less than the set threshold D0, then record the connection point Ci and the point for the same node, adding a new section of R for Ci;

According to the above algorithm, we can get the starting point and the end point of each link by the above algorithm, and get the position attribute of all the connection points and the attributes of link section. The specific attribute information is shown in Tables 1 and 2. Because the location of the map in the cross section is going to do the chain operation, and only take the road of starting point and the stop point for processing, so you can avoid the non-reachable cross of the viaduct and ordinary sections.

Table 1.   Road sections table

No.	Attribute	Physical meaning
1	ID	Road ID
2	Cross1ID	Start point ID
3	Cross2ID	End point ID
4	Length	Road Length
5	Angle	Road Angle

Table 2.   Link point table.

No.	Attribute	Physical meaning
1	ID	Point ID
2	Lng	Longitude
3	Lat	Latitude
4	RoadList	Connecting Road Sections

3.2   The Extraction of Urban Road Network Node

In the process of node extraction of road network, the standardization of GIS data often caused the noise points, therefore, it is required to deal with the noise for the noise nodes. In fact, although the actual crossing of road link number shall be not less than 3, based on the feature, can be deal with the noise for network node information. But for pedestrian bridge, steering section and other sections in some parallel road also have this feature, as shown in Fig. 3, the structure is not included in the scope of the study of backbone topology structure. Therefore, it is necessary to firstly exclude this kind of road section, can do the extraction of road intersection.

Fig. 3. Noises of road node

According to the characteristics of these small sections, the design of denoising algorithm is as follows:

Step 1:   Take all the connection points with linking number 3;

Step 2:   Among the three sections of the connection point A, judge whether two angles are basically the same, connection is basically the same two angles, whether differ greatly from the other angle;

Step 3:   If it is, then judge whether another connection point B, which is different from the other two sections of the road, also has the same characteristics with connection point A;

Step 4:   If it is, then determine the section to be removed sections, excluding sections, modify the list of relevant connection points.

According to the above algorithm, we can get the starting point and the end point of each link by the above algorithm, and get the position attribute of all the connection points and the attributes of link section. The specific attribute information is shown in Tables 1 and 2. Because the location of the map in the cross section is going to do the chain operation, and only take the road of starting point and the stop point for processing, so you can avoid the non reachable cross of the viaduct and ordinary sections.

Extraction algorithm of Network topology node is as follows:

Step 1:   Take all connection points for no less than 3;

Step 2:   Calculating the distance D_{ij} between any two connection points C_i and C_j, if $D_{ij} < D_n$, then C_i and C_j are confirmed to be at the same intersection;

Step 3:   Extract each intersection as Cross group according to the distance relation above;

Step 4:   The average position data of the connection point from each Cross group is taken as the position of the "big node";

Step 5:   Take the set of connection points of each Cross group connection point as link information of "big node".

Step 6:   Number all Cross groups, the combined "big node" is the network topology node.

Through this algorithm, all the connecting points of intersection are converted to the network topology node, the attribute of the road network nodes are expressed as follows (Table 3).

Table 3. Road network node table.

No.	Attribute	Physical meaning
1	ID	Point ID
2	Lng	Longitude
3	Lat	Latitude
4	CrossList	Including connection point
5	RoadList	Connecting Road Sections

3.3   Generation of Topology Link Information

Because the quality and the standard of map data mapping is often uneven, road sections of the road layer are often composed of a number of lines that do not have a unified rule, that is to say, a complete road may be made from a number of sections. And connecting points of these segments, when the connecting point, whose connection degree is 2, comes from the same section, it needs to be consolidated. The connection between topological nodes can be obtained by combining several times that is topology link. Because to add weight to the topology, in the course of the merger must guarantee to generate a topological edge (logical connection), but to record all the connected sections (physical connection). The combination of this section provides the basis for setting the link weights between nodes, and to a certain extent, restore the integrity of the road.

The construction algorithm of topological nodes is as follows:

Step 1:   Obtain the connection relationship of all the connection points;

Step 2:   To merge all the sections connected by the same link with number 2, and remove the connection point;

Step 3:   Perform multiple operations, until all the connection points are intersection point;

Step 4:   According to the relationship between the network node and the connection point, turn the connection point into the road network node, topological edge structure is completed.

The connection relationship of network topology points is calculated by the above algorithm, namely, network topological edge is constructed (logical connection), but link contained in topological edge (physical connection) still needed to be calculated. The construction algorithm for the middle segment of the topological edge is represented as follows:

Step 1:   Get all topological points and topological edges (topological point connection);

Step 2:   Find the link Rt that connected with the link Rs, which starts from the start node of topological edge Ns, add Rt to road section group of topological edge;

Step 3:   If another connection point of Rt is in the end node of topological edge Ne, enter next topological edge to process, otherwise, with Rt as the starting point to continue to find sections, until the complete link is obtained.

Through the above algorithm, the link can be added to any topological edge in actual road network, so attribute of topological edge can be constructed as follows:

As a result, the urban topologic structure can be achieved basically, just need to fill in topological weight in accordance with the attributes of road section, then urban topologic structure with the weight can be constructed.

3.4   Generation of Topology Link Information

Topological edge has been turned into the representation of road section group. In this paper, the average speed of floating car data of Nanning is the speed of the road, the distance from the GIS data is the length of each road, and finally, the travel time of road section group is used as the topological edge weight. The longer travel time, the greater the weight accordingly. That is:

$$W = L/v$$

W means topological edge weights, L means the total length of sections included in the topology, and V means the actual traveling speed obtained from speed of floating cars.

4   System Experiment and Result Analysis

4.1   Laboratory Data Set

The map data used in this paper is the main road data of Nanning City, Guangxi Province, including arterial road and expressway. The geographic information data in the experiment is obtained on the basis of ArcGIS system, in which a total of 12,383 sections of the road, and gained 11,372 corresponding link points.

Floating car data used in the experiment is collected by more than 6 thousand taxis in Nanning city, including the location, direction, speed and other information of the car, the scanning frequency is 10 s–20 s. The collected floating car data and road matched speed as the speed of the matching road. In this experiment, the average value of the speed of the floating car in each road section in 3 min is adopted as the operation speed of the road segment, which is the foundation for weight setting of the link between nodes.

4.2   Acquisition of Nodes

The experiment has selected the topology node for the main road network in Nanning.

It is calculated that we have obtained 11372 road connection points in trunk road network, such as connection point attribute of No. 388 (388, 108.34452739992, 22.7191316400132, [351,353,429]), which means connection point of No. 388 connect to the sections of 351, 353 and 429.

585 junction points are extracted from the connection points of the road, a total of 156 junctions are extracted, and 156 topological nodes are obtained. such as connection point attribute of No. 33 (32,108.30470567,22.834619343, [10004,10033, 10104], [1401319,1402342,1402354,1402355,1402356,15678122,1402344]), namely, the node is connected by 10004, 10033, 10104, and connects the 7 sections as above.

4.3   Extraction of Links

By calculation, 156 topological nodes are finally generated in Nanning road network, with 456 links between nodes, adding weight value to each topology. Such as edge

Fig. 4. Topology of Nanning City.

attributes from node 121 to node 7 (121,7, [91226041,91226041,14560214, 14560219, 14560220, 14560163, 1410048, 1419620], 700, 38.45, 65), which means that edge node 121 can reach the node 7 after the above 8 sections, with the length of 700 meters, the average speed of the road 38.45 km/h, edge weight of 65.

4.4 Generation of Topology Link Information

In this paper, the topology of the backbone network is generated by the NetworkX database, the location information of the original node is preserved when the topology graph is generated, and the generating effect chart is projected on the satellite image, as shown below (Fig. 4).

5 Conclusion

This paper presents an auto-discovery method of dynamic topological structure of urban road network. Firstly, based on the urban geography model, construct the graph model of urban road network, that is, with road network intersection as node and connected path between nodes as links, establish graph model of "node - link" to represent the urban road network topology. Road operation speed information based on floating car data collection adds weights for links, and then construct the weighted directed topology of the urban road network. Taking Nanning as an example, the experimental analysis shows that, this method can be used to find the intrinsic topological characteristics of the complex network in urban areas, achieve dynamic updating of urban road network topology. According to the actual operation state of the road, add weight value to the topology, and provide important topological data support for analyzing the optimal path and urban transport planning.

References

1. Szeto, W., Jiang, Y., Wang, D.: A sustainable road network design problem with land use transportation interaction over time. J. Netw. Spat. Econ. **15**, 791–822 (2015)
2. Jenelius, E., Mattsson, L.-G.: Road network vulnerability analysis of area-covering disruptions: a grid-based approach with case study. J. Transp. Res. Part A Policy Pract. **46**, 746–760 (2012)
3. Tu, Y., Yang, C., Chen, X.: Road network topology vulnerability analysis and application. In: Proceedings of the Institution of Civil Engineers-Transport, pp. 95–104. Thomas Telford Ltd., New York (2013)
4. Banks, N.C., Paini, D.R., Bayliss, K.L., et al.: The role of global trade and transport network topology in the human-mediated dispersal of alien species. J. Ecol. Lett. **18**, 188–199 (2015)
5. Zhang, Q.G., Zhang, L.W., Yin, X.R.: Building of road network topology and the shortest path analysis based on MapX. In: Applied Mechanics and Materials, vol. 380, pp. 2573–2576. Trans Tech Publications (2013)
6. Xu, M., Wu, J., Du, Y.: Discovery of important crossroads in road network using massive taxi trajectories. J. (2014). arXiv preprint: arXiv:1407.2506

7. Karagiorgou, S., Pfoser, D.: On vehicle tracking data-based road network generation. In: Proceedings of the 20th International Conference on Advances in Geographic Information Systems, pp. 89–98. ACM (2012)
8. Mali, G., Michail, P., Paraskevopoulos, A., Zaroliagis, C.: A new dynamic graph structure for large-scale transportation networks. In: Spirakis, P.G., Serna, M. (eds.) CIAC 2013. LNCS, vol. 7878, pp. 312–323. Springer, Heidelberg (2013)
9. Zhang, Z., Virrantaus, K.: Analysis of vulnerability of road networks on the basis of graph topology and related attribute information. In: Phillips-Wren, G., et al. (eds.) Advances in Intelligent Decision Technologies. SIST, vol. 4, pp. 353–363. Springer, Heidelberg (2010)
10. Crisostomi, E., Kirkland, S., Shorten, R.: A Google-like model of road network dynamics and its application to regulation and control. J. Int. J. Control **84**, 633–651 (2011)

Circuit Analysis and Systems

An XDL Analysis Method
for SRAM-Based FPGA

Junfeng Liu, Yunyi Yan[(⊠)], and Jinfu Wu

School of Aerospace Science and Technology,
Xidian University, Xi'an 710071, China
jffliu@163.com, yyyan@xidian.edu.cn

Abstract. Soft errors due to single event upsets (SEUs) are the main challenge to the reliability of SRAM-based FPGA designs. To calculate the soft error rate, the XDL information, such as the sensitive configuration bits, signal propagation, is needed to be acquired. In this paper, an XDL analysis method for SRAM-based FPGA is presented. Firstly, the **inst** information is analyzed to obtain the sensitive configuration bits and the most important logic module LUT. Then, the **net** information is analyzed and the signal propagation in CLBs is presented. Finally, the overall signal propagation of a FPGA design is presented.

Keywords: Xilinx Design Language (XDL) · Field Programmable Gate Array (FPGA) · SRAM · Soft error rate

1 Introduction

Due to its high performance and reconfigurability, Field Programmable Gate Array (FPGA) is widely used in computer hardware, communication and aerospace engineering applications [1, 2]. Compared with the traditional ASIC (Application Specific Integrated Circuit), FPGA has shorter development cycle and faster updates [3]. Especially the Xilinx SRAM-based FPGA is widely used in aerospace engineering application in recent years.

However, the SRAM-based FPGA design is sensitive to single-event upsets (SEUs), due to which its logic can be changed and an error may occur. The error rate due to SEUs is referred to as soft error rate (SER) [4]. To calculate the SER of a FPGA-based design, its information of sensitive configuration bits [4] and the digital circuits [5] is in demand. Xilinx offers a very powerful tool called Xilinx Design Language (XDL) [6], which can be used to describe the circuit layout in a human readable format. With the Xilinx Design Language, the full configuration information, internal routing, and logic structures of any FPGA-based design can be acquired.

This work was supported by the National Natural Science Foundation of China under Grants No. 61571346 and No. 61305041.

J. Pan et al. (eds.), *Intelligent Data Analysis and Applications*, Advances in Intelligent Systems and Computing 535, DOI 10.1007/978-3-319-48499-0_29

Although XDL is designed in a human readable format, it is still not clear enough for those who are not familiar with FPGA structure. Recently, a lot of researchers are focused on analyzing the XDL of FPGA-based designs.

Firstly, the XDL should be created. To achieve it, the conventional Xilinx flow [7] can be used to acquire NCD file at first and then using the command –ncd2xdl [8] to convert NCD to XDL.

With the XDL created Ghosh *et al.* [9] presented that the XDL contains four parts: DESIGN, MODULE, INSTANCE and NET. However, the key parts are only two: INSTANCE and NET. Claus et al. [10] showed how to analyze the INSTANCE and NET inside XDL of Virtex-II. XDL describes the INSTANCE and NET as component **inst** and switch matrix **net**, respectively. The format of the component configuration inside a SLICE of Virtex-II is presented as: **component name::#parameter**. However the format is not accurate. Then there is a detail discuss on **net**. Beckhoff et al. [11] presented a method to analyze for Spartan-6 XDL. A detail of **inst** is presented and the information of LUT is introduced briefly. But how to acquire sensitive configuration bits, digital circuits and other more information is not introduced both in [9–11].

So far, there exists no literature related to the XDL information analysis of FPGA Virtex-4 and Virtex-5. In addition, the differences between Virtex-4 and Virtex-5 designs in terms of CLB routing, SLICE number, and instances description in the XDL should be studied in detail. Only to find out a general XDL analytical method for different kinds of device types, the SER calculation of FPGA-based designs can be designed with good robust.

In this paper, a XDL analysis method including **inst** and **net** explanations is presented. Firstly, the **inst** information is analyzed to obtain the sensitive configuration bits related to the SER and the most important logic module LUT. Then, the **net** information is analyzed and the signal propagation in CLBs is presented. Finally, the overall signal propagation of a multiplier FPGA design is presented using the **inst** and **net** information analysis.

The remainder of this paper is organized as follows. Section 2 describes the proposed XDL analysis method. Section 3 presents the experimental results. Finally, Sect. 4 concludes the paper.

2 Proposed XDL Analysis Method

It is obvious that the XDL of Virtex-4 and Virtex-5 comprise two parts: **inst** and **net**. **Inst** information describes the detail configurations of instance, while **net** is the physical connections between these instances. The **inst** and *net* information is necessary for building practical circuits and calculating the soft error rate.

2.1 Analysis of Inst

For a FPGA design, the most basic functional blocks are I/O block (IOB), SLICE and BRAM [2]. I/O block determines the inputs or outputs of a design, while SLICE includes Look-Up Table (LUT) which presents the logic of signals. BRAM is the memory element, whose sensitive configuration bits can be obtained with the similar way of I/O and SLICE. For simplicity, only the I/O block and SLICEL are analyzed in the following.

2.1.1   I/O Block

(a) *Virtex-4.*

I/O block is the input or output of SRAM-based FPGA. Once an error occurs in IOB, the function of SRAM-based FPGA is likely to be changed. Therefore, it is necessary to analyze the IOB in XDL when calculating the SER. The **inst** information representing IOB is described as follows.

```
inst "y""IOB",placed CIOB_X17Y29 R1   ,
   cfg " DIFFI_INUSED::#OFF DIFF_TERM::#OFF IMUX::#OFF
        OUSED::0 PADOUTUSED::#OFF PULLTYPE::#OFF
        TUSED::#OFF OUTBUF:y_OBUF: PAD:y: DRIVE::12
        OSTANDARD::LVCMOS25  SLEW::SLOW "
   ;
```

The syntax for **inst** information is: **inst**"name""type", **placed** "tile""site", **cfg** "string". In the example above, the name of the instance is **y** and the type of the instance is **IOB**. The components are described after the key word **cfg**. The description format for components in Virtex-4 FPGA is **component::parameter**. For example, **IMUX::#OFF** indicates that the component is unused, where **IMUX** and **#OFF** denotes the name and parameter of the component, respectively. All the components are presented in **inst** and they are all sensitive configuration bits. In addition, **OUT-BUF:y_OBUF** represents that instance **y** is an output instance. If **OUTBUF** is configured as **INBUF**, i.e., **OUTBUF: INBUF**, instance **y** is an input instance.

(b) *Virtex-5.*

Virtex-5 has the same basic IOB diagram as Vietex-4. But there are some differences in the description of the XDL. The configuration of an output instance in Virtex-5 is shown below:

```
inst "y""IOB",placed CIOB_X17Y29 R1   ,
   cfg " DIFFI_INUSED::#OFF DIFF_TERM::#OFF IMUX::#OFF
        OUSED::0 PADOUTUSED::#OFF PULLTYPE::#OFF
        TUSED::#OFF OUTBUF:y_OBUF: PAD:y: DRIVE::12
        OSTANDARD::LVCMOS25  SLEW::SLOW "
   ;
```

The method to distinguish between input instance and output instance is the same with Virtex-4. But the sensitive configurations bits are different in IOB between them.

2.1.2 SLICE

(a) *Virtex-4*

The simplified general SLICE of Virtex-4 is shown in Fig. 1. Virtex-4 FPGA function generators are implemented as 4-input look-up tables (LUTs). There are two LUTs (F and G) in a SLICE. And each LUT has four independent inputs. For an example of SLICE in the XDL of Virtex-4:

```
inst "y_OBUF""SLICEL",placed CLB_X33Y78 SLICE_X47Y157  ,
  cfg "...
        DXMUX::#OFF DYMUX::#OFF F::#OFF F5USED::#OFF
        ...
        FXMUX::#OFF FXUSED::#OFF G:y1:#LUT:D=(A3*A2)
        ..."
;
```

When calculating SER, the LUTs inside the SLICEL are required. In the example above, the LUT of function generators F is empty and (A3*A2) is the function generators G. The A1, A2, A3, A4 correspond to the each four inputs of function generator F and G. Signal input into the SLICEL, then through the LUT (F or G), finally exit SLICEL from the output port.

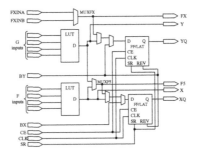

Fig. 1. Simplified Virtex-4 FPGA general SLICE

(b) *Virtex-5*

The components and routing inside SLICEL in Virtex-5 is shown in Fig. 2. It indicates that every SLICEL contain four 6-input look-up tables (A, B, C and D). Signal enter into look-up tables from the port A1 ~ A6, B1 ~ B6, C1 ~ C6 and D1 ~ D6. There are two exit ports (O5 and O6) for signal to choose to exit each LUT.

Then how the instance is described in XDL will be discussed. For example:

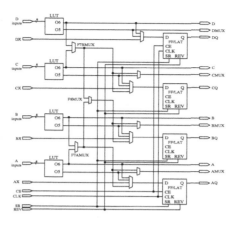

Fig. 2. Diagram of SLICE inside Virtex-5 FPGA

```
inst "y_OBUF""SLICEL",placed CLBLL_X16Y29 SLICE_X26Y29 ,
  cfg "A5LUT::#OFFA6LUT:y1:#LUT:O6=(A3*A6) ACY0::#OFF
        AFF::#OFF AFFINIT::#OFF AFFMUX::#OFF AFFSR::#OFF
        AOUTMUX::#OFF AUSED::0 B5LUT::#OFF B6LUT::#OFF
        ...
        REVUSED::#OFF SRUSED::#OFF SYNC_ATTR::#OFF "
  ;
```

The form to describe component in Virtex-5 is also **component::parameter**. Inside the XDL, there are two described LUT for one function generator. For example, the two LUT in function generator A are **A5LUT** and **A6LUT**. Inside the contents of **A6LUT** the flag **O6** shows that signals through the **A6LUT** and exit from the port **O6**. And the LUT in **A6LUT** is **(A3*A6)**.

Table 1 summarizes the differences between Virtex-4 and Virtex-5 in the sensitive configuration bits of SLICE.

Table 1. Sensitive configuration bits of SLICE

Devices	Components of SLICE
Virtex-4	BXINV, BYINV, CEINV, CLKINV, COUTUSED, CY0F, CY0G, CYINIT, DXMUX, DYMUX, F, F5USED, FFX, FFX_INIT_ATTR, FFX_SR_ATTR, FFY, FFY_INIT_ATTR, FFY_SR_ATTR, FXMUX, FXUSED, G, GYMUX, REVUSED, SRINV, SYNC_ATTR, XBUSED, XMUXUSED, XUSED, YBUSED, YMUXUSED, YUSED
Virtex-5	A5LUT, A6LUT, AFF, AFFINIT, AFFMUX, AFFSR, AOUTMUX, AUSED, B5LUT, B6LUT, BCY0, BFF, BFFINIT, BFFMUX, BFFSR, BOUTMUX, BUSED, C5LUT, C6LUT, CCY0, CEUSED, CFF, CFFINIT, CFFMUX, CFFSR, CLKINV, COUTMUX, COUTUSED, CUSED, D5LUT, D6LUT, DCY0, DFF, DFFINIT, DFFMUX, DFFSR, DOUTMUX, DUSED, PRECYINIT, REVUSED, SRUSED, SYNC_ATTR

2.2   Analysis of Net

(a) *Virtex-4*

Net is also a very important part of the XDL, which presents the connections between instances inside a FPGA design. There is an example of Virtax-4 design below:

```
net "a_IBUF" ,
  outpin "a" I ,
  inpin "y_OBUF" G3 ,
  pip CLB_X33Y78 IMUX_B22_INT -> G3_PINWIRE3 ,
  pip INT_X33Y78 OMUX_WS1 -> IMUX_B22 ,
  pip INT_X34Y79 BEST_LOGIC_OUTS2 -> OMUX1 ,

  pip IOIS_LC_X34Y79 IOIS_I1 -> BEST_LOGIC_OUTS2_INT ,
  pip IOIS_LC_X34Y79 IOIS_IBUF1 -> IOIS_IBUF_PINWIRE1 ,
  pip IOIS_LC_X34Y79 IOIS_IBUF_PINWIRE1 -> IOIS_I1 ,
;
```

Net is key word for the begin of the **net** and it is followed by the name of the **net**. In our example the **net** is called **a_IBUF**. Then the **outpin** and **inpin** is presented and always have the form: **outpin "instance" output port and inpin "instance" input port**. In simple terms, **outpin** is the begin instance of the **net** and the **inpin** is the end instance of the **net**. In other words, the signals output from the **outpin** and input to the **inpin**. In the example above, the signals output from the instance **a** and then input to the instance **y_OBUF**. And the port of the **begin** and **end** is also presented: the signals output from the port **I** of **a** and input to **y_OBUF** from the port **G3**.

Inside **net**, the **pip** shows the connection between the CLBs and the form of the pip is always **pip tile wire0 → wire1**. There are four tiles in our example: **CLB_X33Y78, IOIS_LC_X34Y79, INT_X33Y78 and INT_X34Y79**. Details of the connection are shown below (Fig. 3):

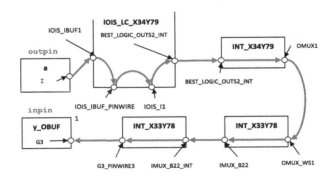

Fig. 3. Brief description of PIP

(b) *Virtex-5*

The same method can be used to analyze the *net* of Virtex-5. Inside the **net**, the CLBs (**tile** in **pip**) and the ports of tiles(**outpin** and **input**) are different from Virtex-4 (but the vhd file is same). The **net** of Virtex-5 is listed follow:

```
net "a_IBUF" ,
    outpin "a" I ,
    inpin "y_OBUF" A3 ,
    pip CLBLL_X16Y29 SITE_IMUX_B27 -> M_A3 ,
. . .
    ;
```

Within the analysis of **net**, how the signal propagates is clear. Further, combine with **inst**, the digital circuit is acquired in Sect. 3.

3   Experimental Results

With XDL, the overall signal propagation in a FPGA design, including instances and the connections between those instances can be obtained. For each instance, its name, type, number of sensitive configuration bits (Bits), input port, output port and logic (LUT), which are crucial for the SER calculation, can be acquired from the **inst** information in XDL. Figure 4 is the analysis results of a multiplier FPGA design based on Virtex-4.

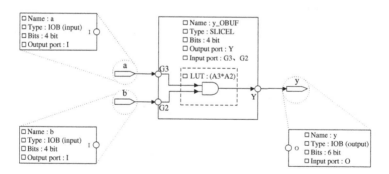

Fig. 4. Analysis results

4   Conclusion and Future Work

Soft errors due to SEUs are the main reliability threat of digital systems. To calculate soft error rate, the sensitive configuration bits, signal propagation and more must to be known. The XDL is helpful to achieve it. Combining with **inst** and **net** information inside XDL, the necessary information for SER's calculation can be obtained. In this paper, **inst** and **net** information is analyzed by our proposed method and the information is shown in analysis results.

Potential future work in the area includes, e.g. exploring a more comprehensive analysis method for other configuration information, which can achieve more accurate SER calculation for the FPGA-based design.

References

1. Srinivasan, S., Gayasen, A., Vijaykrishnan, N.: Improving soft-error tolerance of FPGA configuration bits. In: International Conference on Computer-Aided Design, pp. 107–110. IEEE (2004)
2. Zhu, M., Song, N., Pan, X.: Mitigation and experiment on neutron induced single-event upsets in SRAM-based FPGAs. IEEE Trans. Nucl. Sci. **60**(4), 3063–3073 (2013)
3. Zarandi, H.R., Miremadi, S.G., Pradhan, D.K.: Soft error mitigation in switch modules of SRAM-based FPGAs. In: IEEE International Symposium on Circuits and Systems, pp. 141–144. IEEE (2007)
4. Asadi, G., Tahoori, M.B.: Soft error rate estimation and mitigation for SRAM-based FPGAs. In: Proceedings of the 2005 ACM/SIGDA 13th International Symposium on Field-Programmable Gate Arrays, pp. 149–160. ACM (2005)
5. Asadi, G., Tahoori, M.B.: An accurate ser estimation method based on propagation probability. Comput. Sci. **1**, 306–307 (2005)
6. Wang, X.F, Si, S.H., Gao, C.: A method of FPGA interconnect resources testing by using XDL-based configuration. In: Prognostics and System Health Management Conference, pp. 203–207. IEEE (2014)
7. Lavin, C., Padilla, M., Lundrigan, P.: Rapid prototyping tools for FPGA designs: RapidSmith. In: 2010 International Conference on Field-Programmable Technology (FPT), pp. 353–356. IEEE (2010)
8. Cheremisinov, D.I.: Design automation tool to generate EDIF and VHDL descriptions of circuit by extraction of FPGA configuration. In: East-West Design & Test Symposium, pp. 1–4. IEEE (2013)
9. Ghosh, S., Nelson, B.: XDL-based module generators for rapid FPGA design implementation. In: International Conference on Field Programmable Logic and Applications, FPL 2011, 5–7 September, China, Crete, Greece, pp. 64–69 (2011)
10. Claus, C., Zhang, B., Hübner, M.: An XDL-based busmacro generator for customizable communication interfaces for dynamically and partially reconfigurable systems. In: Works on Reconfigurable Computing Education (2007)
11. Beckhoff, C., Koch, D., Torresen, J.: The xilinx design language (XDL): tutorial and use cases. In: International Workshop on Reconfigurable Communication-Centric Systems-On-Chip, pp. 1–8 (2011)

Area Estimation for Triple Modular Redundancy Field Programmable Gate Arrays

Hongjie He[(✉)], Baolong Guo, and Yunyi Yan

School of Aerospace Science and Technology,
Xidian University, Xi'an 710071, China
jiehh_2008@163.com, blguo@xidian.edu.cn

Abstract. SRAM-based FPGAsare more vulnerable to single-event, particularly the single-event upset (SEU). Triple modular redundancy (TMR) protection method is a good method to prevent the FPGA system from being destroyed. At the same time, the area cost of an FPGA project plays an important role in designing an FPGA system. What's more, the protection method can enlarge the area cost, it may exceed the maximum of the FPGAs. So we should know the changes of area when the project is protected. This article firstly discusses a new way to estimate the area of an FPGA project, which is based on its used resources, such as Look-Up-Table (LUT), Brams, IO and so on. Then, we offer the model to estimate the new area cost of the project when it's protected by TMR. At last, we apply TMR Tool to verify the models and the results are sound.

Keywords: Field programmable gate arrays (FPGA) · Area · Used resources · Triple modular redundancy (TMR)

1 Introduction

The area cost of an FPGA project is an important index to evaluate the total cost, so it's worth studying on it. FPGAs, such as Xilinx Virtex, are a class of programmable devices which use static random access memory(SRAM)cells for implementing logic and so on [1]. In space environment, single event upset(SEU) induced by cosmic ray in electronic devices, and the SRAM cells are highly susceptible to SEU, which seriously imperils the safety of spacecraft [2]. So protection of the devices is needed and its cost estimation is critical, which influences the evaluation of the protective methods. However, when we take some ways to protect the FPGAs, we can analysis the changed number of specific used resources, so we seek a new way to estimate the FPGA project's area cost roughly. To an FPGA project, the used resource includes Look-Up-Table (LUT), Brams, IO and so on, and in our work, we will build a model to estimate the area with the used resources.

There are many protective methods to against SEU [3], such as triple modular redundancy protection method (TMR), refresh cycle protection method and hamming

This work was supported by the National Natural Science Foundation of China under Grants No. 61571346 and No. 61305041.

code protection method, and TMR is an effective one [4]. So, in this paper, we mainly study the TMR method and the project's area cost when it's protected by this method and we will build the area cost models of it. When the electronic devices works in space, SEU can make some bits flip, TMR can complete tolerance backup and correct the mistakes, which will prevent the errors traveling downward [5]. In Fig. 1, we divide the protected code into M modules and each module has N bits. The bit can be same or not to different modules. To Module 1, it has N1 bits, we make redundancy backup it into two pieces, and named B1-1, B1-2 and B1-3. So the three pieces have the same number of bit, what's more, they are mutual independent. We also divide the performance period, and regard a clock cycle as basic unit. We suppose that the SEU may occur on any bit in every clock cycle, the error happens on one period will not get into next clock circle because of the protection of TMR. When give the specific clock, the sub-modules of Module i, Bi-1, Bi-2 and Bi-3 have the same Ni bit. To a certain sub-module Ni bits, one or more bits occurs upset called sub-module's error, if other sub-modules stay right, TMR can avoid system's error.

We build the area cost model of TMR based on the changed used resources, so we can know the area when the projects are protected by TMR easily. To a specific model FPGA, its area is limited, so we can judge the module can be protected by TMR or not. At last we use TMRTool, which is developed by Xilinx and used to build a triplication scheme into digital designs, to verify the area model of TMR.

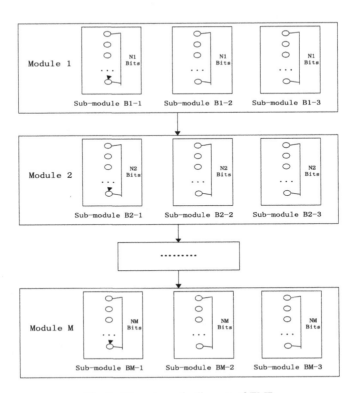

Fig. 1. The schematic diagram of TMR

2   The Models

2.1   The Area Model of an FPGA Project

Many related works have been done, such as Siew-Kei Lam and Thambipillai Srikanthan proposed estimation technique relies on a novel approach to partition the custom instruction data-paths into a set of clusters, here each cluster can be realized using an FPGA logic element or a coarse-grained arithmetic unit [6]. However, this method did not consider Brams' effect on area, which influences a lot. For an FPGA module, area consumption refers to the usage of hardware resources. It not only contains the number of used LUT, Bram and so on, but also layout resources. In this work, according to the quantity of used resources, we can estimate the area cost. What's more, as we can infer the changing of used resources when the project is protected by TMR, so it helps building the Post-TMP project's area cost model. The consumption of area can be defined as follows:

$$S = \sum_{i=1}^{n} \alpha_i \cdot s_i \tag{1}$$

where, S presents the total area consumption, α_i for the normalized weighting factor and s_i for the area consumption.

In this work, we build area model mainly focus on the bits of LUT, Slice, Brams, IO and their corresponding routing resource that occupy. For LUT, we just consider their configuration bit, for Slice and IO, we just think about their control bit, however, for Brams, we not only discuss its configuration bit, but also control bit.

LUT Area. As we all know, XC4VSX55 FPGA has four-input LUTS [7], so each LUT can be regarded as a RAM that has four address lines and we can set its normalized weighting factor as $2^4 = 16$.

SLICE Area. For XC4VSX55, each configurable logic blocks(CLBs) comprises four logic slices, and each logic slice has two 4-LUTs and two flip-flops. According to the composition of each CLB, we know the control bit of SLICE is 32, consequently, we set the normalized weighting factor of SLICE as 32 [8].

Bram Area. For XC4VSX55, we learn that its configuration bit is 16 K, and its control bit is 21and we set its weighting factor as 21.

IO Area. In our work, we think each IO includes two bits, they control the input and output respectively, so we set its weighting factor as 2.

Routing Area. Base on the design experience and related information, routing area is about four times as large as the sum of the LUT area, Slices area and a part of Bram area. So, the routing area can be expressed as follows:

$$Area_{rout} = 4 * (\alpha * N_{lut} + \beta * N_{bram} + \gamma * N_{slice\_used} + \lambda * N_{io}) \tag{2}$$

where, $Area_{rout}$ indicates the area of routing. $\alpha = 16$, $\beta = 21$, $\gamma = 2$.

So, the area can be expressed in Eq. (3):

$$Area = 5 * (\alpha * N_{lut} + \beta * N_{bram} + \gamma * N_{slice\_used} + \lambda * N_{io}) + \eta * N_{bram} \qquad (3)$$

where, *Area* indicates the area of an FPGA project, $\eta = 16 * 1024 = 16386$. For a given FPGA project, the main step is to obtain its used resources. We get the number of Brams, LUTS, Slices and IO from the file of *.mrp(an example in Fig. 2), which is generated from the ISE(Integrated Software Environment), designed by Xilinx.

```
Design Summary
--------------

Design Summary:
Number of errors:       0
Number of warnings:    32
Logic Utilization:
    Number of Slice Flip Flops:       1,906 out of  49,152      3%
    Number of 4 input LUTs:           2,054 out of  49,152      4%
Logic Distribution:
    Number of occupied Slices:                      1,645 out of  24,576     6%
        Number of Slices containing only related logic:   1,645 out of   1,645   100%
        Number of Slices containing unrelated logic:          0 out of   1,645    0%
        *See NOTES below for an explanation of the effects of unrelated logic
Total Number of 4 input LUTs:         2,224 out of  49,152      4%
    Number used as logic:             2,054
    Number used as a route-thru:        170
    Number of bonded IOBs:               28 out of     640      4%
    Number of BUFG/BUFGCTRLs:             2 out of      32      6%
        Number used as BUFGs:             2
        Number used as BUFGCTRLs:         0
    Number of FIFO16/RAMB16s:             7 out of     320      2%
        Number used as FIFO16s:           0
        Number used as RAMB16s:           7
    Number of DSP48s:                    14 out of     512      2%

Total equivalent gate count for design:  489,938
Additional JTAG gate count for IOBs:  1,344
Peak Memory Usage:   307 MB
Total REAL time to MAP completion:    11 secs
Total CPU time to MAP completion:     11 secs
```

Fig. 2. Illustration of used resources in *.mrp file

From Fig. 2, we can get the number of used resources easily, Brams is 7, LUT is 2224, Slices is 1645 and IO is 28. According Eq. (3), we have calculated 20 FPGA projects' area, and the results are showed in Table 1.

2.2 The Area Model of TMR

Based on TMR method [9], we take redundancy on FPGA projects, which need be protected. Then, we select majority by resolvers for fault tolerance and to prevent the errors.

To the area, compared with primitive used resources, we think, in theory, the number of LUT, Bram and Slice becomes there times as many as before when the module is protected by TMR and the number of IO stay the same. What's more,

Table 1. 20 FPGA projects' used resources and their respective area

Number	LUT	Brams	IO	Slices	Area
1	1187	4	44	928	309492
2	72	1	40	55	31131
3	1349	5	44	954	343103
4	5404	32	108	3702	1552568
5	5979	15	121	3265	1248327
6	8995	9	95	5311	1717969
7	29680	104	131	18540	7056126
8	2115	8	45	1947	612738
9	7784	24	76	5219	1853696
10	678	2	23	372	146788
11	6291	13	70	4268	1400683
12	5489	10	60	3583	1177430
13	1354	5	90	972	346475
14	546	0	22	410	109324
15	1705	5	43	1381	439901
16	2691	30	53	1250	910116
17	3120	21	63	2130	936837
18	7016	61	71	5680	2476173
19	447	0	21	233	73082
20	247	2	14	138	74850

because of the increasement of the used resources, the using placement and routing increase much.

TMR can bring extra area, which is caused by the AND-OR gates and the placement and routing. So, we should add this part, and it is set as experience point. According to our related work we set the value as 0.4 times as large as the primitive area.

$$S_p \approx [5 * (3 * \alpha * N_{LUT} + 3 * \beta * N_{BRAM} + 3 * \gamma * N_{Slice\_used}) + 3 * \eta * N_{BRAM} + \lambda * N_{io}] + s_x \quad (4)$$

where, S_p represents the area of the modules, which are protected by TMR, s_x is the area of AND-OR gates and the placement and routing and its value is 0.3* *Area*.

We calculate the TMR area with Eq. (4), and the results are shown in Table 2.

3   The Validation

On the base of the cost model of an FPGA project, we offer the area model of TMR, the model is built according to our theoretical analysis, so we should verify them. Here we use TMRTool, by which we can get the real number of used resources, and then we can

Table 2. The same 20 FPGA projects' used resources and their area, which are protected by TMR

Number	TMR-LUT	TMR-BRAMs	TMR-IO	TMR-SLICE	TMR-Area
1	3561	12	44	2784	1052096
2	216	3	40	165	105685
3	4047	15	44	2862	1166374
4	16212	96	108	11106	5278299
5	17937	45	121	9795	4243827
6	26985	27	95	15933	5840714
7	89040	312	131	55620	23990304
8	6345	24	45	5841	2083129
9	23352	72	76	15657	6302262
10	2034	6	23	1116	498987
11	18873	39	70	12804	4762042
12	16467	30	60	10749	4003022
13	4062	15	90	2916	1177655
14	1638	0	22	1230	371613
15	5115	15	43	4143	1495491
16	8073	90	53	3750	3094182
17	9360	63	63	6390	3184993
18	21048	183	71	17040	8418704
19	1341	0	21	699	248394
20	741	6	14	414	254434

calculate the FPGA project's area, which is protected by TMR. After that, we compare the real result with the result that we get the TMR area model.

The detailed steps are as follows [10]:

3.1 The Steps

Step 1: Create an ISE project to implement the pre-XTMR design. The FPGA project's file (*.vhd) can generate *.ngc file and *.mrp file by Integrated Software Environment (ISE). We get the number of BRAMS, LUTS, I/O and DSP from the file of *.mrp.

Step 2: Create a TMRTool project and generate the XTMR netlist. We get the *. ngc file and put it to TMR tool.

Step 3: Create a second ISE project to implement the XTMR design. We can get a *.edif file through last step, then we put the file into ISE and get a *mrp file again, just like the first step above, and obtain the number of used resources and with our models we can calculate the power and area which are Post-XTMP (being protected by TMR).

3.2   The Results

As we mentioned above, TMR Tool is a good tool for our validation. We can get the changed used resources information with its help, so we can calculate the new area value with the Eq. (3) and then get Table 3.

Table 3. The projects'used resources get by TMRTool and the model's error

Number	TMR-LUT	TMR-Brams	TMR-IO	TMR-Slices	POST TMR-Area	TMR-Area	Error
1	3724	13	44	2830	1058038	1052096	0.56 %
2	225	4	40	169	120423	105685	12.24 %
3	4052	15	44	2875	1134543	1166374	2.81 %
4	16231	96	108	11121	5126962	5278299	2.95 %
5	17946	46	121	9798	4136686	4243827	2.59 %
6	26992	28	95	15941	5687248	5840714	2.70 %
7	89051	312	131	55701	23298531	23990304	2.97 %
8	6351	25	45	5863	2042346	2083129	1.99 %
9	23371	72	76	15661	6119052	6302262	2.99 %
10	2039	6	23	1119	485188	498987	2.84 %
11	18883	39	70	12818	4625013	4762042	2.96 %
12	16471	30	60	10752	3886079	4003022	3.00 %
13	4071	15	90	2921	1144527	1177655	2.89 %
14	1643	0	22	1237	362201	371613	2.60 %
15	5116	15	43	4149	1452541	1495491	2.95 %
16	8080	90	53	3760	3005330	3094182	2.95 %
17	9366	64	63	6393	3108761	3184993	2.45 %
18	21051	185	71	17056	8206868	8418704	2.58 %
19	1346	0	21	701	241806	248394	2.72 %
20	745	7	14	419	264560	254434	3.82 %

Table 3 shows the 20 projects' used resources, which is got by TMRTool and we calculate the area with Eq. (3). In Table 3, "POST TMR-Area" refers to the area that we calculate by the used resources that we get by TMRTool, and we regard it as the true value, and we use it verifying the TMR area model.

From the table, we find the average error is about 3.18 %, which we think the model, which we have built in Sect. 2.2 is good.

4   Conclusion

In this paper, we, firstly, study a good method to estimate an FPGA project's area, which is based on its used resources and then, on this base, we get the area model of TMR by our theoretical analysis. At last, we verify the model by TMRTool, and the result is sound.

References

1. Legat, U., Biasizzo, A., Novak, F.: SEU recovery mechanism for SRAM-based FPGAs. IEEE Trans. Nucl. Sci. **59**(5), 2562–2571(2012)
2. Bidokhti, N.: SEU concept to reality (allocation, prediction, mitigation). In: Reliability & Maintainability Symposium, pp. 1–5. IEEE (2010)
3. Single-Event Upset Mitigation for Xilinx FPGA Block Memories. http://www.xilinx.com/support/documentation/application_notes/xapp962.pdf
4. Samudrala, P.K., Ramos, J., Katkoori, S.: Selective triple modular redundancy (STMR) based single-event upset (SEU) tolerant synthesis for FPGAs. IEEE Trans. Nucl. Sci. **51**(5), 2957–2969(2014)
5. Matsumoto, K., Uehara, M., Mori, H.: Stateful TMR for transient faults, world automation congress (WAC), pp. 1–6. IEEE (2010)
6. Lam, S.K., Srikanthan, T.: Estimating area costs of custom instructions for FPGA-based Reconfigurable Processors. J. 89–94 (2007)
7. XinlinxVirtex-4SeriesFPGAs. http://china.xilinx.com/publications/matrix/virtex4_color.pdf
8. Virtex-4 FPGA Configuration User Guide, Xilinx Inc.UG071, ver. 1.10 (2008)
9. Quinn, H., Morgan, K., Graham, P., et al.: Domain crossing errors: limitations on single device triple-modular redundancy circuits in xilinx FPGAs. IEEE Trans. Nucl. Sci. **54**(6), 2037–2043 (2008)
10. Xilinx TRMTool User Guide: TMRTool Software Version 9.2i, Xilinx Inc, UG156, ver. 2.2 (2009)

Coupled Plasmonic Nanoantennas

Hancong Wang[1,2,3(✉)]

[1] Fujian Provincial Key Laboratory for Automotive Electronics
and Electric Drive, Fuzhou 350108, China
whcuser@163.com
[2] Fujian Provincial Key Laboratory of Digital Equipment,
Fuzhou 350108, China
[3] School of Information Science and Engineering,
Fujian University of Technology, Fuzhou 350108, China

Abstract. The electromagnetic coupling between metal nanoparticles lead to a variety of fundamental studies and practical applications in plasmonics. For example, by strong coupling between metallic nanostructures, plasmonic antennas are able to concentrate and re-emit light in a controllable way. A variety of structures of optical antennas have been investigated in the past decade. In this review, we will discuss the coupled plasmonic nanoantennas from the typical applications point of view, i.e., control of local intensity, control of emission direction, control of far-field polarization, and outlook the corresponding impacts in understanding physics.

Keywords: Plasmonics · Nanoantennas · Coupling · Generalized Mie theory

1 Introduction

Antennas appeared a century ago was first used to transmit and collect radio and microwave which now play an essential role in the modern wireless world [1–4]. Optical antennas as an analogue at the nanoscale are of great interest due to the unique ability of controlling absorption and emission at visible and infrared region [4, 5], such as focusing optical fields to sub-diffraction limited volumes [6–8], enhance the excitation and emission of molecules [9, 10] and quantum emitters [11, 12] and modify their spectrum and lifetime [13, 14]. Propagating light can be converted into nanoscale enhanced near field [15–18], and vice versa, a localized excitation can be coupled to directed radiation. The efficiency of an optical antenna depends on its shape, material, dimension, geometry, and operation frequency [19, 20]. Various optical antennas have been developed experimentally and theoretically, such as individual discs [21], triangles [22, 23], flowers [24], as well as coupled antenna such as dimers [3, 25], bowties [8, 13], trimers [10], etc. [26–28]. Most of these antennas are based on metallic nanostructures that support surface plasmons (SP) [29, 30], so called plasmonic antennas. Single and coupled nanoantennas have been investigated thoroughly by far-field spectroscopy, exploiting two-photon luminescence [9, 31] and near-field scanning microscope [4, 22, 32]. Abundant applications have been found in surface enhanced Raman scattering [6, 33], optical manipulation [25, 34], biosensing [35, 36] and integrated photonic devices [37–39], etc.

© Springer International Publishing AG 2017
J. Pan et al. (eds.), *Intelligent Data Analysis and Applications*, Advances in Intelligent Systems and Computing 535, DOI 10.1007/978-3-319-48499-0_31

Here we provide an overview of the coupled antenna system, e.g. coupled nanorod, bow-tie and nano-aggregates, etc. Light properties such as the intensity, polarization and direction of emission, can be controlled effectively by different geometries of nanoantennas, and will be discussed respectively.

2   Control of Local Intensity

To enhance the performance of nanoantennas, the way of plasmonic coupling between nanostructures are usually adopted. Figure 1a shows two nanorod with the length 500 nm separated about 40 nm [9]. The coupling between these two rods gives rise to strong local enhancement which is detected by the intensity of TPL. More experiments and theoretical calculations indicate that, the intensity enhancement at the visible frequencies can be as high as 10^3 when the gap reduced to a few of nanometers.

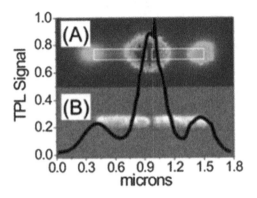

Fig. 1. (Reproduced from [9]) (a) TPL scans for two 500 nm long gold bar with a 40 nm gap and (b) the SEM image: The superimposed black lines plot the TPL signal along the symmetry axis of the antennas.

The bow-tie structures [13] are one of the most investigated configurations. The enhancement factor of $|E|^2/|E_0|^2$ at the gap of the bow-tie antenna is evaluated by the surface-enhanced fluorescence which is shown in Fig. 2. The size of the gap varies from 15 nm to 80 nm. The maximum enhancement of fluorescence f_F can reach to 1340-fold.

More systematic study was done by Schnell et al., where they investigated the near-field oscillations of progressively loaded plasmonic antennas at infrared frequency by scattering-type scanning near-field optical microscopy [4]. In Fig. 3a the amplitude signals at the both ends of the nanorod and phase change at the rod center, clearly reveals the dipolar mode excited at a continuous nanorod. When a wedge is cut at the middle, the dipolar mode of the nanorod still holds which is shown in Fig. 3b. But, if the rod is cut more deeply and even fully cut, the case is completely different. In Fig. 3c, the bridge between the two rods is only 2 % of the cross-section, which cannot restore the dipolar mode any more. For the fully cut rod in Fig. 3d, a gap (80 nm) is formed in the rod center. Each antenna segment oscillates as a dipole. Hence, a face change exists between the two segments and the gap.

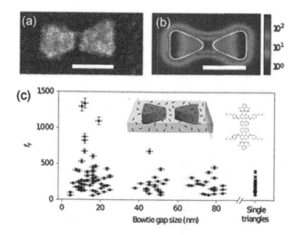

Fig. 2. (Reproduced from [13]) (a) SEM image of a gold bowtie antenna. Scale bar = 100 nm. (b) Calculation of the local electric field intensity enhancement. (c) Enhancement factor, f_F, from several nanoantennas as function of the gap size. Inset: Schematic representation of molecules randomly placed around a gold bowtie antenna on a transparent substrate.

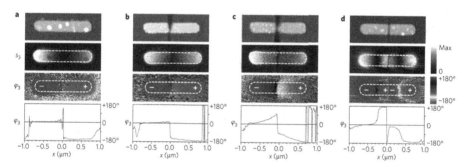

Fig. 3. (Reproduced from [4]) Near-field images of progressively loaded nanoantennas at a wavelength of 9.6 m. (a) Continuous rod antenna. (b) Low-impedance loaded antenna where a thick metal bridge connects the two antenna segments. (c) High-impedance loaded antenna where a tiny metal bridge connects the two antenna segments. (d) Fully cut antenna where the two antenna segments are completely separated. Experimental results showing topography and near-field amplitude and phase images.

3   Control of Emission Direction

Controlling the far-field emission from an emitter is another important property of nanoantennas. For an emitter such as quantum dot, the emission direction involves a large solid angle. To beaming the emission, a Yagi-Uda plasmonic antenna are developed [11]. In Fig. 4, a quantum dot is positioned near one of the arm of the antenna. The constructive interference of the emission from each arm excited by the emitter results in a narrow directional radiation pattern. The gap between each arm in Yagi-Uda is about 100 nm, which is relative large for plasmonic coupling.

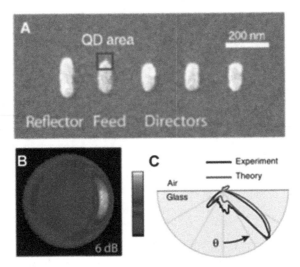

Fig. 4. (Reproduced from [11]) (a) SEM image of a five-element Yugi-Uda gold nanoantenna. A quantum dot is attached to one end of an arm, indicated with a red square. (b) Radiation pattern (intensity distribution at the back focal plane of the objective) from Yagi-Uda after an 830-nm long-pass filter. (c) Measurement (black line) and calculation (red line) of the radiation angular distribution for the structure. (Color figure online)

A stronger coupling case is shown in Fig. 5 [40]. The scattering properties can be tuned by two closely spaced silver and gold disks. Interestingly, the direction of the scattering is dependent on the wavelength. For blue light at 450 nm and red light at 700 nm, the scattering is in opposite directions shown in Fig. 5b, which can be used as an ultra-small $\sim\lambda/100$ photon-sorting nanodevice.

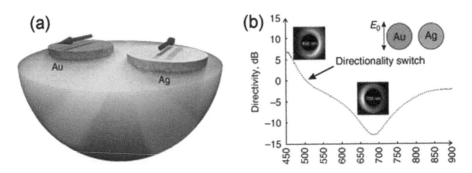

Fig. 5. (Reproduced from [40]) (a) An artist's view of colour routing from a bimetallic dimer supported by a glass substrate. (b) Experimental directivities as a function of wavelength. Insets: corresponding direct and Fourier color images. Radiation patterns recorded at specific wavelengths corresponding to scattering to the right (450 nm) and to the left (700 nm). (Color figure online)

4   Control of Far-Field Polarization

Not only the intensity and direction, but also light polarization from an emitter such as Raman scattering (RS) of molecules in the gap between nanoantennas can be manipulated significantly [10, 41]. First a simplest case of a nanocrystal dimer is considered. Figure 6a shows the SEM image of such a dimer, and it is seen that the angle of the dimer axis, is $\sim 130°$ with respect to the x axis. This angle is corresponding to the maximal RS intensity shown in the Fig. 6b by black and red dots, which means the favorite incident polarization of the laser for the enhancement is along the dimer axis for both Raman shift (773 cm^{-1} and 1650 cm^{-1}). Here the low concentration of molecules used ensures that each aggregate contains no more than a single molecule [42]. The depolarization ρ in Fig. 6c is defined as $\rho = (I_{//} + I_{\perp})/(I_{//} + I_{\perp})$, where $I_{//}$ and $I_{\perp}$ are RS signals with orthogonal polarization, which indicates the polarization of the RS light is also along the dimer axis at these two frequencies. These properties of dimer can be simulated by treating the nanoparticles as spheres. As the single molecule Raman signal can only from "hot sites", the calculations are concentrated on the field enhancement in the junction of the dimer at the laser wavelength, and the depolarization of the emission from a dipole situated in this junction. The calculated results shown by the curves in Fig. 6b and c are consistent to the experiment. This expected agreement is actually due to the axial symmetric geometry of the dimer.

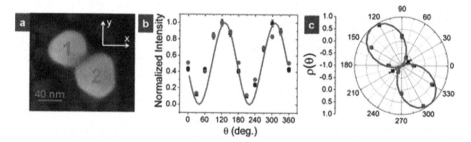

Fig. 6. (Reproduced from [10]) Polarization response of a nanocrystal dimer. (a) A SEM image shows a dimer of nanoparticles, which is tilted ~ 1300 from the x direction. (b) Normalized SERS intensity at 555 nm (black squares) and 583 nm (red circles) as a function of the angle of rotation by the λ/2 waveplate. The green line is the result of the generalized Mie theory calculation of the normalized local field enhancement factor at $\lambda = 532$ nm, using the geometry from the SEM image as the only input. (c) Depolarization ratio (ρ) measured at 555 nm (back squares) and 583 nm (red circles). Black and red lines show the result of Mie theory calculation performed at 555 nm and 583 nm, respectively. (Color figure online)

A drastically different behavior is obtained from the trimer shown in Fig. 7. The intensity profile in Fig. 7b is maximal at an angle of $\sim 75°$, which is close to the axis of 1st and 2nd nanoparticles. Whereas, the depolarization ratio profiles do not coincide with each other and in addition they are both rotated with respect to the intensity profile. The depolarization pattern of the 555 nm light is rotated by $\sim 45°$, while the 583 nm light is rotated by $\sim 75°$. In order to simulate this situation of trimer, calculation was performed by assuming that the molecule is placed in turn in each of the three possible junctions. Only when the molecule is set in the junction marked with a red arrow in Fig. 7a, the calculated and experimental results are in good agreement for both normalized intensity and depolarization, which also confirms the assumption that only one molecule in the junction, contributes to the signal. What should be note is this counter-intuitive wavelength-dependent polarization rotation is not an accident. The rotation only exists in the cases with the number of the particles are larger than two.

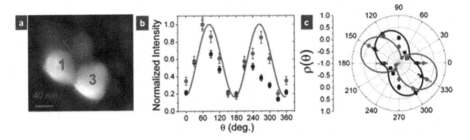

Fig. 7. (Reproduced from [10]) Polarization response of a nanocrystal trimer. (a) SEM image of the trimer. (b) Normalized SERS intensity at 555 nm (black squares) and 583 nm (red circles) as a function of the angle of rotation by the λ/2 waveplate. The intensities at both wavelengths show approximately the same profile, but the maximal intensity is observed at ~75°, which does not match any pair of nanoparticles in the trimer. The green line the result of a calculation assuming that the molecule is situated at the junction marked with red arrow in SEM image. (c) Depolarization ratio (ρ) measured at 555 nm (back squares) and 583 nm (red circles). The black and red lines show the result of calculations at the two wavelength, assuming that the molecule is situated at the junction marked with red arrow in SEM image. (Color figure online)

5   Summary

Various coupled nanoantennas have been introduced here. Plasmonic antennas can be used to manipulate light properties at the nanoscale. Compare with single structures, more important nanoantennas are coupled systems, such as nanorod with small gap, bow-tie, Yugi-Uda structure and nanoparticle aggregates, etc. With the help of SP coupling, light intensity, direction and polarization can be well tailed. Although there are still a lot of unsolved problems, it is no doubt, nanoantennas as a new subject will be further developed, and more applications will be found in, such as ultra-sensitive sensor, and biosensing, integrated photonic devices, etc.

Acknowledgements. This work was supported by the Startup Foundation of Fujian University of Technology (GY-Z160049), the Mid-youth Project of Education Bureau of Fujian Province (JAT160331), and the Fujian Provincial Major Research and Development Platform for the Technology of Numerical Control Equipment (2014H2002).

References

1. Greffet, J.J.: Nanoantennas for light emission. Science **308**, 1561–1563 (2005)
2. Mühlschlegel, P., Eisler, H.J., Martin, O.J.F., Hecht, B., Pohl, D.W.: Resonant optical antennas. Science **308**, 1607–1609 (2005)
3. Shegai, T., Miljkovic, V.D., Bao, K., Xu, H.X., Nordlander, P., Johansson, P., Kall, M.: Unidirectional broadband light emission from supported plasmonic nanowires. Nano Lett. **11**, 706–711 (2011)
4. Schnell, M., Garcia-Etxarri, A., Huber, A.J., Crozier, K., Aizpurua, J., Hillenbrand, R.: Controlling the near-field oscillations of loaded plasmonic nanoantennas. Nat. Photonics **3**, 287–291 (2009)
5. Giannini, V., Fernandez-Dominguez, A.I., Heck, S.C., Maier, S.A.: Plasmonic nanoantennas: fundamentals and their use in controlling the radiative properties of nanoemitters. Chem. Rev. **111**, 3888–3912 (2011)
6. Xu, H.X., Bjerneld, E.J., Käll, M., Börjesson, L.: Spectroscopy of single hemoglobin molecules by surface enhanced ranman scattering. Phys. Rev. Lett. **83**, 4357–4360 (1999)
7. Zhang, Z., Weber-Bargioni, A., Wu, S.W., Dhuey, S., Cabrini, S., Schuck, P.J.: Manipulating nanoscale light fields with the asymmetric bowtie nano-colorsorter. Nano Lett. **9**, 4505–4509 (2009)
8. Fromm, D.P., Sundaramurthy, A., Schuck, P.J., Kino, G., Moerner, W.E.: Gap-dependent optical coupling of single "Bowtie" nanoantennas resonant in the visible. Nano Lett. **4**, 957–961 (2004)
9. Ghenuche, P., Cherukulappurath, S., Taminiau, T.H., van Hulst, N.F., Quidant, R.: Spectroscopic mode mapping of resonant plasmon nanoantennas. Phys. Rev. Lett. **101**, 116805 (2008)
10. Shegai, T., Li, Z.P., Dadosh, T., Zhang, Z., Xu, H.X., Haran, G.: Managing light polarization via plasmon–molecule interactions within an asymmetric metal nanoparticle trimer. Proc. Nat. Acad. Sci. U.S.A. **105**, 16448–16453 (2008)
11. Curto, A.G., Volpe, G., Taminiau, T.H., Kreuzer, M.P., Quidant, R., van Hulst, N.F.: Unidirectional emission of a quantum dot coupled to a nanoantenna. Science **329**, 930–933 (2010)
12. Ringler, M., Schwemer, A., Wunderlich, M., Nichtl, A., Kurzinger, K., Klar, T.A., Feldmann, J.: Shaping emission spectra of fluorescent molecules with single plasmonic nanoresonators. Phys. Rev. Lett. **100**, 203002 (2008)
13. Kinkhabwala, A., Yu, Z.F., Fan, S.H., Avlasevich, Y., Mullen, K., Moerner, W.E.: Large single-molecule fluorescence enhancements produced by a bowtie nanoantenna. Nat. Photonics **3**, 654–657 (2009)
14. Farahani, J.N., Pohl, D.W., Eisler, H.J., Hecht, B.: Single quantum dot coupled to a scanning optical antenna: a tunable superemitter. Phys. Rev. Lett. **95**, 017402 (2005)
15. Knight, M.W., Grady, N.K., Bardhan, R., Hao, F., Nordlander, P., Halas, N.J.: Nanoparticle-mediated coupling of light into a nanowire. Nano Lett. **7**, 2346–2350 (2007)
16. Li, Z.P., Hao, F., Huang, Y.Z., Fang, Y.R., Nordlander, P., Xu, H.X.: Directional light emission from propagating surface plasmons of silver nanowires. Nano Lett. **9**, 4383–4386 (2009)
17. Li, Z.P., Zhang, S.P., Halas, N.J., Nordlander, P., Xu, H.X.: Coherent modulation of propagating plasmons in silver-nanowire-based structures. Small (Weinheim an der Bergstrasse, Germany) **7**, 593–596 (2011)

18. Li, Z.P., Bao, K., Fang, Y.R., Guan, Z.Q., Halas, N.J., Nordlander, P., Xu, H.X.: Effect of a proximal substrate on plasmon propagation in silver nanowires. Phys. Rev. B **82**, 241402 (2010)
19. Wiley, B.J., Chen, Y.C., McLellan, J.M., Xiong, Y.J., Li, Z.Y., Ginger, D., Xia, Y.N.: Synthesis and optical properties of silver nanobars and nanorice. Nano Lett. **7**, 1032–1036 (2007)
20. Liang, H.Y., Yang, H.X., Wang, W.Z., Li, J.Q., Xu, H.X.: High-yield uniform synthesis and microstructure-determination of rice-shaped silver nanocrystals. J. Am. Chem. Soc. **131**, 6068–6069 (2009)
21. Langhammer, C., Kasemo, B., Zoric, I.: Absorption and scattering of light by Pt, Pd, Ag, and Au nanodisks: absolute cross sections and branching ratios. J. Chem. Phys. **126**, 194702–194711 (2007)
22. Rang, M., Jones, A.C., Zhou, F., Li, Z.Y., Wiley, B.J., Xia, Y.N., Raschke, M.B.: Optical near-field mapping of plasmonic nanoprisms. Nano Lett. **8**, 3357–3363 (2008)
23. Nelayah, J., Kociak, M., Stephan, O., de Abajo, F.J.G., Tence, M., Henrard, L., Taverna, D., Pastoriza-Santos, I., Liz-Marzan, L.M., Colliex, C.: Mapping surface plasmons on a single metallic nanoparticle. Nat. Phys. **3**, 348–353 (2007)
24. Liang, H.Y., Li, Z.P., Wang, W.Z., Wu, Y.S., Xu, H.X.: Highly surface-roughened "Flower-like" silver nanoparticles for extremely sensitive substrates of surface-enhanced raman scattering. Adv. Mater. **21**, 4614–4618 (2009)
25. Svedberg, F., Li, Z.P., Xu, H.X., käll, M.: Creating hot nanoparticle pairs for surface-enhanced raman spectroscopy through optical manipulation. Nano Lett. **6**, 2639–2641 (2006)
26. Jin, R.C., Cao, Y.W., Mirkin, C.A., Kelly, K.L., Schatz, G.C., Zheng, J.G.: Photoinduced conversion of silver nanospheres to nanoprisms. Science **294**, 1901–1903 (2001)
27. Aizpurua, J., Hanarp, P., Sutherland, D.S., Kall, M., Bryant, G.W., de Abajo, F.J.G.: Optical properties of gold nanorings. Phys. Rev. Lett. **90**, 057401 (2003)
28. Hao, F., Nehl, C.L., Hafner, J.H., Nordlander, P.: Plasmon resonances of a gold nanostar. Nano Lett. **7**, 729–732 (2007)
29. Raether, H.H.: Surface Plasmons. Springer (1988)
30. Lee, K.G., Kihm, H.W., Kihm, J.E., Choi, W.J., Kim, H., Ropers, C., Park, D.J., Yoon, Y.C., Choi, S.B., Woo, H., Kim, J., Lee, B., Park, Q.H., Lienau, C., Kim, D.S.: Vector field microscopic imaging of light. Nat. Photonics **1**, 53–56 (2007)
31. Schuck, P.J., Fromm, D.P., Sundaramurthy, A., Kino, G.S., Moerner, W.E.: Improving the mismatch between light and nanoscale objects with gold bowtie nanoantennas. Phys. Rev. Lett. **94**, 017402 (2005)
32. Taminiau, T.H., Moerland, R.J., Segerink, F.B., Kuipers, L., van Hulst, N.F.: lambda/4 Resonance of an optical monopole antenna probed by single molecule fluorescence. Nano Lett. **7**, 28–33 (2007)
33. Wang, W., Li, Z.P., Gu, B.H., Zhang, Z.Y., Xu, H.X.: Ag@SiO2 core-shell nanoparticles for probing spatial distribution of electromagnetic field enhancement via surface-enhanced raman scattering. ACS Nano **3**, 3493–3496 (2009)
34. Li, Z.P., Käll, M., Xu, H.: Optical forces on interacting plasmonic nanoparticles in a focused Gaussian beam. Phys. Rev. B **77**, 085412 (2008)
35. Lal, S., Clare, S.E., Halas, N.J.: Nanoshell-enabled photothermal cancer therapy: impending clinical impact. Acc. Chem. Res. **41**, 1842–1851 (2008)
36. Loo, C., Lowery, A., Halas, N.J., West, J., Drezek, R.: Immunotargeted nanoshells for integrated cancer imaging and therapy. Nano Lett. **5**, 709–711 (2005)
37. Ozbay, E.: Plasmonics: merging photonics and electronics at nanoscale dimensions. Science **311**, 189–193 (2006)

38. Kirchain, R., Kimerling, L.: A roadmap for nanophotonics. Nat. Photonics **1**, 303–305 (2007)
39. García de Abajo, F.J., Cordon, J., Corso, M., Schiller, F., Ortega, J.E.: Lateral engineering of surface states - towards surface-state nanoelectronics. Nanoscale **2**, 717–721 (2010)
40. Shegai, T., Chen, S., Miljkovic, V.D., Zengin, G., Johansson, P., Kall, M.: A bimetallic nanoantenna for directional colour routing. Nat. Commun. **2**, 2749–2763 (2011)
41. Li, Z.P., Shegai, T., Haran, G., Xu, H.X.: Multiple-particle nanoantennas for enormous enhancement and polarization control of light emission. ACS Nano **3**, 637–642 (2009)
42. Le Ru, E.C., Meyer, M., Etchegoin, P.G.: Proof of single-mokeule sensitivity in surface enhanced Raman scattering (SERS) by means of a two-analyte technique. J. Phys. Chem. B **110**, 1944–1948 (2006)

A High Frequency Voltage-Controlled PWM/PSM Dual-Mode Buck DC-DC Converter

Zhong Lun-Gui[1(✉)] and Cheng Xin[2]

[1] Fujian University of Technology, Fuzhou 350118, China
zlg@fjut.edu.cn
[2] School of Electronic Science and Applied Physics,
Hefei University of Technology, Hefei 230009, China

Abstract. A high frequency voltage-controlled PWM/PSM dual-mode Buck DC-DC converter is presented in this paper. In order to achieve high efficiency over its entire load range, pulse-width modulation (PWM) and pulse-skip modulation (PSM) were integrated in the proposed Buck converter. By adopting a novel load current sensing circuit, the converter's operation mode can be automatically changed between PWM and PSM mode according to load current. Different from the traditional inductor current sensing circuit, the proposed load current sensing circuit is realized by detecting the gate driver signal of synchronous power NMOS which changes with the load current simultaneously. The proposed Buck converter has been designed and simulated in SMIC 0.18 μm CMOS process. Simulation results show that the peak conversion efficiency of the Buck converter is about 85 %. It is able to regulate the output voltage at 1.2 V from a 1.6 ~ 3 V supply voltage for the maximum output current of 500 mA with 20 MHz switching frequency.

1 Introduction

With the development of semiconductor industry and wireless communication, DC-DC converter has been widely used in portable devices, such as smart phones, laptops, PDA and so on. High frequency and high power efficiency are two major design requirements of Buck regulators in portable devices [1]. Firstly, the value of inductor and capacitor can be reduced significantly when the switching frequency of the converter is raised. The high frequency Buck converter can integrate both inductor and capacitor in a single chip, and can reduce the number of external components [2]. Secondly, the power consumption of these feature-rich electronic products is highly increased, while the battery technology is difficult to have a big breakthrough in short period of time, so how to improve conversion efficiency becomes the key problem in the design of DC-DC converter [3].

The power consumption of DC-DC converter can be divided into three parts: conduction loss, switching loss and quiescent dissipation [4]. Traditional PWM mode have the advantages of high efficiency, low ripple and high driving capability at heavy loads. However, the efficiency will decreases drastically at light load since the

© Springer International Publishing AG 2017
J. Pan et al. (eds.), *Intelligent Data Analysis and Applications*, Advances in Intelligent Systems and Computing 535, DOI 10.1007/978-3-319-48499-0_32

switching loss dominates [5]. PSM mode is a new control scheme that maintains high efficiency at light load by minimizing the switching frequency thus reduce switching loss. Furthermore, PSM mode has the advantages of better EMI performance than PFM mode [6]. In order to increase the conversion efficiency over a wide operation current range, PWM mode and PSM mode can be combined together on a converter chip [7].

In this paper, a high frequency voltage-controlled PWM/PSM dual-mode Buck DC-DC converter is presented. The converter's operation mode can be automatically changed between PWM and PSM mode according to load current by adopting a novel load current sensing circuit. Different from the traditional inductor current sensing circuit, the proposed load current sensing circuit is realized by detecting the gate driver signal of synchronous MOSFET which change with the load current simultaneously. The presentation of this paper is organized as follows. Section 2 introduces the converter system structure and its compensation method. The circuit implementation is shown in Sect. 3. Section 4 presents the simulation results. Finally, conclusion is made in Sect. 5.

2   System Structure

Figure 1 shows the system block diagram of the proposed dual-mode Buck converter. The buck converter contains two power transistors, one LC filter, and dual-mode control circuit. The main dual-mode control circuit consists of PWM mode circuit, PSM mode circuit, PWM/PSM selector circuit, zero current sensing circuit, oscillator, soft start circuit, logic control circuit, etc.

The working principle can be described as following. Firstly, after power up, the oscillator (OSC) generates a nearly 20-MHz clock signal CLK as the input signal for the RS flip-flop which will be used in PSM mode and a saw-tooth signal V_{saw} for the PWM comparator which will be used in PWM mode. A reference voltage V_{ref} is generated by the voltage reference block. In order to avoid output overshoot current, a soft start block is adopted to make sure the output voltage V_{refs} rise to its final value V_{ref} linearly.

In PWM mode, the EA amplifies the error between the feedback voltage V_{fb} and V_{refs} to generate the error signal V_c. Then the PWM comparator will compare the output voltage V_c with the ramp voltage V_{ramp} to generate the PWM mode control signal V_{PWM}. In PSM mode, the PSM comparator compares the feedback voltage V_{fb} with V_{refs} to generate the signal V_e, then the signal V_e and the clock signal CLK are sent to RS flip-flop to generate the PSM mode control signal V_{PSM}. The PSM/PWM mode selector is a key circuit of the dual-mode converter proposed in this paper. It can realize automatic change between PWM and PSM mode according to load current.

Besides, the dead-time control circuit is designed to avoid the shoot through current when the power PMOS and power NMOS are turned on simultaneously. While the zero current detector (ZCD) can shut down power NMOS when zero current occurs to avoid power consumption.

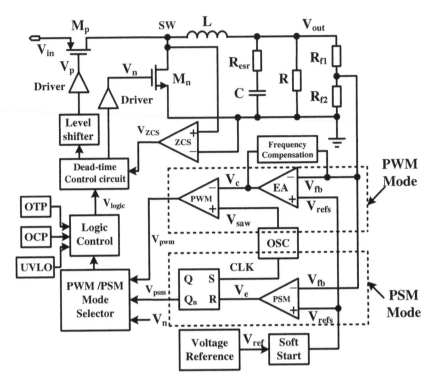

Fig. 1. System block diagram of the proposed dual-mode buck converter.

3   Circuit Implementation

3.1   Ramp Generator

The proposed Buck converter requires an on-chip ramp generator to generate the ramp signal [11]. As shown in Fig. 2(a), the ramp generator comprises a capacitor, a charging branch consist of transistor M_7 and M_8, a discharging current branch consist of transistor M_6 and M_9, a comparator and an R-S flip-flop. The two comparators compare V_{ramp} with the upper voltage band VH and lower voltage band VL to provide V_{ramp} for the R-S flip-flop and generate the clock signal CLK. The timing diagram of the ramp generator is shown in Fig. 2(b). When CLK is low, M_8 is turned on, the capacitor C_1 is charged by the current I_1. On the contrary, when CLK is high, M_9 turns on, the capacitor C_1 discharges to ground.

3.2   PWM Comparator

The main function of PWM comparator is to compare the output voltage V_c of the error amplifier with the ramp voltage V_{ramp} for generating PWM modulator signal [12]. The PWM comparator is shown in Fig. 3 which is formed by a source-coupled differential pair with positive feedback to provide a high DC gain. The DC gain of the PWM comparator can be given by

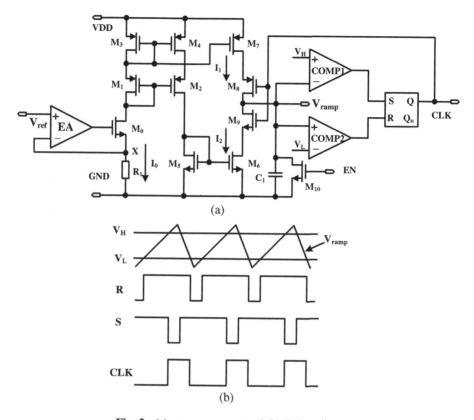

Fig. 2. (a) ramp generator and (b) timing diagram

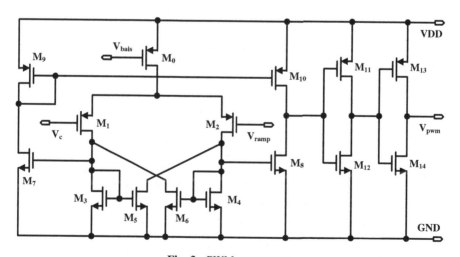

Fig. 3. PWM comparator

$$A_v = \frac{g_{m1}g_{m8}}{g_{m3}}(r_{08} \parallel r_{10})\frac{1}{1-\alpha} \tag{2}$$

Where $\alpha = (W/L)_5/(W/L)_3$ is the positive feedback factor. The inverter chains $M_{11} \sim M_{14}$ are mainly used to increase the driving ability and response speed of the comparator output signal.

3.3 Modulator Mode

When decreasing the load current, the converter automatically switches into PFM mode in which the power stage operates intermittently, based on load demand. Due to a reduced switching activity at power stage in PSM mode, the switching losses are minimized, and the device runs with a minimum quiescent current and maintains high efficiency. The block diagram of a simplified PFM mode buck converter is illustrated in Fig. 4. It contains a comparator and an R-S flip-flop. When $V_{fb} > V_{refs}$, V_e goes down, and V_{psm} outputs high voltage no matter CLK is high or low. That is to say, several periods of CLK will be skipped by PSM modulator circuit. When $V_{fb} < V_{refs}$ and CLK is high, V_{psm} holds the previous value, therefore, V_{psm} goes down when CLK is low.

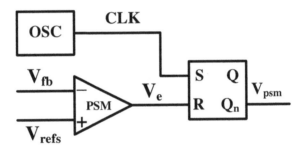

Fig. 4. PSM modulator mode

3.4 PWM/PSM Mode Selector

Figure 5(a) shows the relationship between power NMOS gate control signal V_n and the load current I_{out}. The gate driver signal V_n of synchronous power NMOS can change with the load current simultaneously. The PWM/PSM mode selector is shown in Fig. 5(b). It contains a load current sensing circuit and a multiplexer (MUX). The main load current sensing circuit consists of an RC low-pass filter and comparator. Different from the traditional inductor current sensing circuit, the proposed load current sensing circuit is realized by detecting the gate driver signal V_n of synchronous power NMOS which changes with the load current simultaneously. V_L is the output voltage of the filter which will change with the load current simultaneously, too. The relationship between load current and V_L can expressed as

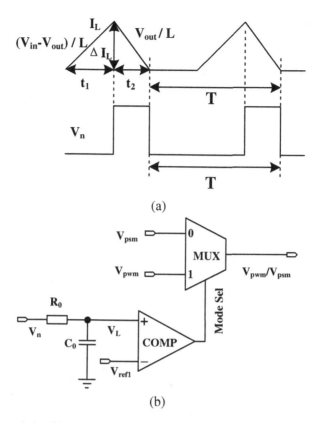

(a)

(b)

Fig. 5. (a) The relationship between power NMOS gate control signal and the load current. (b) PWM/PSM mode selector circuit.

$$I_{out} = \frac{1}{2Lf_{clk}} V_L^2 \frac{V_{out}}{V_{in}} \frac{1}{V_{in} - V_{out}} \tag{3}$$

Where V_{out}, V_{in} and I_{out} are the output voltage, input voltage and load current of the proposed Buck converter, respectively, f_{clk} is the switching frequency of the buck converter, L is the value of the filter inductor.

By comparing the voltage V_L with the reference voltage V_{ref1}, a logic signal V_{pwm} or V_{psm} is generated at output of mode selector circuit. If $V_L > V_{ref1}$, Mode Sel goes up, high level will be selected to send to the multiplexer, then, the output signal of mode selector circuit will be V_{pwm}. The proposed Buck converter will be operated on PWM mode. Contrarily, if $V_L < V_{ref1}$, Mode Sel goes down, low level will be selected to send to the multiplexer, then, the output signal of mode selector circuit will be V_{psm}. The proposed Buck converter will be operated on PSM mode.

3.5   Dead-Time Control Circuit

If both the power PMOS and NMOS are turned on during the transitions, the inductor current will reverse to cause extra power consumption [13–15]. Hence, a dead-time control circuit shown in Fig. 6 is used to avoid the occurrence of this situation. The V_{ZCS} is the output signal of zero current detecting circuit which is also used as a control

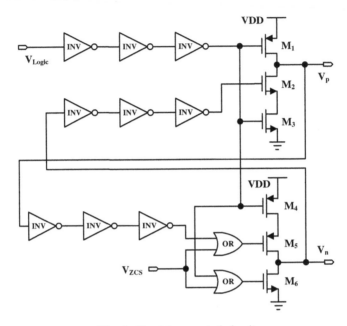

Fig. 6. Dead-time control circuit

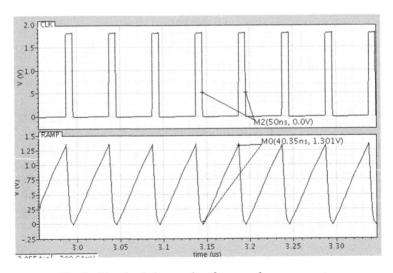

Fig. 7. The simulation results of proposed ramp generator

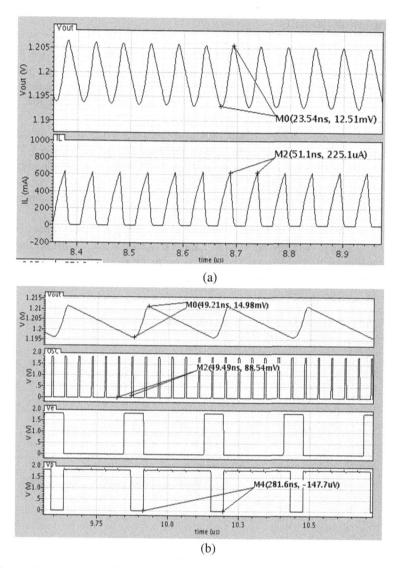

Fig. 8. Steady state simulated results of the converter. (a) PWM mode simulated waveform at I_{out} = 200 mA. (b) PSM mode test waveform at I_{out} = 50 mA.

input signal of this control circuit. When V_{ZCS} goes up, both the power PMOS and NMOS are turned off to prevent the inductor current reversing.

4   Simulation Results

The proposed voltage-controlled PWM/PSM dual-mode Buck converter has been designed and simulated in SMIC 0.18 μm CMOS process. Figure 7 shows plots of the ramp generator voltages V_{ramp} and clock signal CLK, The voltage amplitude of the ramp generator signal V_{ramp} is 1.3 V, and the frequency of clock signal CLK is 20 MHz.

Simulation results under different load conditions are given in Fig. 8. All of the simulations are carried out under the same voltage conditions of V_{in} = 1.8 V, V_{out} = 1.2 V. From Fig. 8(a), the converter operates in PWM mode when the load current is 200 mA, and the ripple voltage is about 12.5 mV. From Fig. 8(b), the converter operates in PSM mode when the load current is 50 mA, and the ripple voltage is about 15 mV.

Figure 9 shows the line transient response of the proposed Buck converter. When the supply voltage is switched between 3 V and 1.6 V within 100 ns, the settling time is 3 μs and the undershoot voltage is 55 mV. The settling time is 2.5 μs and the overshoot voltage is 50 mV when the supply voltage changes from 3 V to 1.6 V within 100 ns.

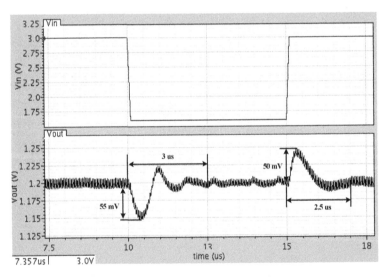

Fig. 9. The line transient responses of the proposed Buck converter

The load transition simulation result between I_{out} = 500 mA and I_{out} = 20 mA is given in Fig. 10, which shows that the converter realizes automatic mode switching smoothly, and the switching time is less than 4 μs. The maximum ripple voltage in the switching process is less than 450 mV.

The conversion efficiency curve is shown in Fig. 11, which shows that the peak efficiency of the high frequency dual-mode converter proposed in this paper is about 85 % at I_{out} = 400 mA and the efficiency is higher than 55 % when I_{out} is larger than 20 mA.

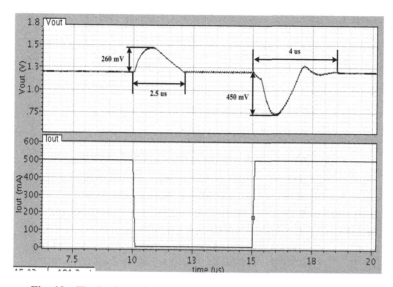

Fig. 10. The load transient responses of the proposed Buck converter

A summary of the proposed LDO regulator performance is shown in Table 1. The peak conversion efficiency of the Buck converter is about 85 %. It is able to regulate the output voltage at 1.2 V from a 1.6 ~ 3 V supply voltage for the maximum output current of 500 mA with 20 MHz switching frequency.

The performances of the proposed Buck converter are summarized in Table 2 with comparison to some recently reported DC-DC converter. As can be seen, the proposed converter achieves a high efficiency when the clock frequency is high. Also, the output voltage ripple is the smallest.

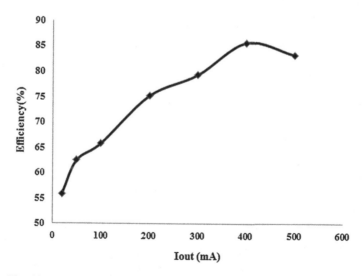

Fig. 11. The conversion efficiency curve of the proposed Buck converter

Table 1. Performance summary of the proposed dual-mode Buck converter

Supply voltage	1.6–3 V	Output voltage	1.2 V
Switching frequency	20 MHz	Peak efficiency	85 %
Voltage ripple	<15 mV	Max. output current	500 mA
Load regulation	0.89 %/A	Line regulation	0.14 %/V
Modulator mode	PWM/PSM	Mode switching point	Iout = 100 mA

Table 2. Summary and compare of performance

	[1]	[2]	[5]	[6]	This work
Years	2011	2012	2011	2014	2015
Process (μm)	0.35	0.13	0.35	0.18	0.18
Freq. (MHz)	5	50	1	3	20
Vin (V)	2.7–4.2	3.3/2.5	3.2–4	2.8–5.5	1.6–3
Vout (V)	0.5–2.9	2.0/1.8	0.8–3.2	1.2	1.2
Ripple (mV)	<40	300	<40	<40	<15
Peak efficiency	91 %	76.8 %	88.5 %	93 %	85 %
Load current	500 mA	300 mA	450 mA	1200 mA	500 mA
Line regulation	10.5 mV/V	7.14 %/V	0.38 %/V	NA	0.14 %/V
Load regulation	8 mV/mA	0.046 %/mA	1.2 ppm/mA	NA	0.89 %/A

5   Conclusion

A high frequency voltage-controlled PWM/PSM dual-mode Buck DC-DC converter is presented in this paper. In order to achieve high efficiency over its entire load range, PWM mode and PSM mode were integrated in the proposed Buck converter. By adopting a novel load current sensing circuit, the Buck's operation mode can be automatically changed between PWM with PSM mode according to load current. The peak conversion efficiency of the Buck converter is about 85 % with 20 MHz switching frequency.

Acknowledgments. This paper is supported by "National Natural Science Foundation of China" (No. 61401137) & "Startup Fund for Scholars of Fujian University of Technology" (No. GY-Z11002).

References

1. Du, M., Lee, H., Liu, J.: A 5-MHz 91 % peak-power-efficiency buck regulator with auto-selectable peak-and valley-current control. IEEE J. Solid-State Circ. **46**(8), 1928–1939 (2011)
2. Ahn, Y., Nam, H., Roh, J.: A 50-MHz fully integrated low-swing buck converter using packaging inductors. IEEE Trans. Power Electron. **27**(10), 4347–4356 (2012)

3. Dongpo, C., Lenian, H., Xiaolang, Y.: A 750 mA, dual-mode PWM/PFM step-down DC–DC converter with high efficiency. J. Semicond. **29**(8), 1614 (2008)
4. Xinquan, L., Huali, Z., Qiang, Y., Huisen, H., Shasha, Z., Yuqing, S.: Design of high efficiency dual-mode buck DC-DC converter. J. Semicond. **31**(11), 115005 (2010)
5. Hwang, B.H., Jhang, Y.C., Chen, J.J., Hwang, Y.S.: A dual-mode fast-transient average-current-mode buck converter without slope-compensation. Microelectron. J. **42**(2), 291–298 (2011)
6. Cheng, J., Ma, Z., Zhang, H.: A voltage mode buck DC–DC converter with automatic PWM/PSM mode switching by detecting the transient inductor current. Analog Integr. Circ. Sig. Process. **80**(2), 243–253 (2014)
7. Chunhong, Z., Haigang, Y., Shi, R.: A wide load range, multi-mode synchronous buck DC-DC converter with a dynamic mode controller and adaptive slope compensation. J. Semicond. **34**(6), 065003 (2013)
8. Chen, J.J., Hsu, J.H., Hwang, Y.S., Yu, C.C.: A DC–DC buck converter with load-regulation improvement using dual-path-feedback techniques. Analog Integr. Circ. Sig. Process. **79**(1), 149–159 (2014)
9. Liou, W.R., Yeh, M.L., Kuo, Y.L.: A high efficiency dual-mode buck converter IC for portable applications. IEEE Trans. Power Electron. **23**(2), 667–677 (2008)
10. Miao, Y., Baixue, Z., Yun, C., Fengfeng, S., Weifeng, S.: A voltage-mode DC—DC buck converter with fast output voltage-tracking speed and wide output voltage range. J. Semicond. **35**(5), 055005 (2014)
11. Wang, C.C., Chen, C.L., Sung, G.N., Wang, C.L.: A high-efficiency DC–DC buck converter for sub-2 × VDD power supply. Microelectron. J. **42**(5), 709–717 (2011)
12. Lee, C.S., Ko, H.H., Kim, N.S.: Integrated current-mode DC–DC boost converter with high-performance control circuit. Analog Integr. Circ. Sig. Process. **80**(1), 105–112 (2014)
13. Lee, Y.T., Wei, C.L., Chen, C.H.: An integrated step-down DC-DC converter with low output voltage ripple. In: 2010 The 5th IEEE Conference on Industrial Electronics and Applications (ICIEA), pp. 1373–1378. IEEE, June 2010
14. Chen, J.J., Shen, P.N., Hwang, Y.S.: A high-efficiency positive buck–boost converter with mode-select circuit and feed-forward techniques. IEEE Trans. Power Electron. **28**(9), 4240–4247 (2013)
15. Liu, P.J., Ye, W.S., Tai, J.N., Chen, H.S., Chen, J.H., Chen, Y.J.: A high-efficiency CMOS DC-DC converter with 9-μs transient recovery time. IEEE Trans. Circ. Syst. I: Regular Pap. **59**(3), 575–583 (2012)

A Complex Network Based Classification of Covered Conductors Faults Detection

Tomas Vantuch[1(✉)], Jan Gaura[1], Stanislav Misak[2], and Ivan Zelinka[1]

[1] Department of Computer Science, VSB-Technical University of Ostrava,
17. listopadu 15, 708 33 Ostrava-Poruba, Czech Republic
{tomas.vantuch,jan.gaura,ivan.zelinka}@vsb.cz
[2] Department of Electrical Power Engineering, VSB-Technical University of Ostrava,
17. listopadu 15, 708 33 Ostrava-Poruba, Czech Republic
stanislav.misak@vsb.cz

Abstract. Presence of partial discharges implies the fault behavior on insulation system of medium voltage overhead lines, especially with covered conductors (CC). This paper covers the machine learning model based on features, which are derived from complex networks. These features are applied to predict whether the measured signal contains phenomenon indicating CC fault behavior or not. The comparison of different threshold levels of similarity values brings more information about complex network modeling. The final performance of the Random Forest classification algorithm shows valuable results for future research.

Keywords: Partial discharges · Complex networks · Random forest · Feature extraction · Covered conductors

1 Introduction

The medium voltage (MV) overhead lines with covered conductors (hereinafter CC) represents the special kind of conductors with additional insulation system which implies their higher operational reliability and safety in comparison to the conductors of Aluminium-Conductor Steel-Reinforced (ARCS) type [17]. This is the reason why CC can be placed on a post in smaller interphase distance and built up area is smaller as well. In case of interphased touch of individual conductors of CC or the contact of with branches of a tree does not arise interphase short-circuit. This is a main advantage of MV overhead lines with CC. For this advantage CC are mostly installed in the forested and broken terrain.

However, in case of CC rupture with subsequent downfall of the line, this fault is not possible to detect by standard digital relays because the earth fault does not arise [6]. The low-energy current passes through the fault point, which implies that standard digital relays working on current principle cannot detect this fault. Nevertheless, it is possible to detect a partial discharge (PD) activity in fault point, which generates inhomogeneous electric field around a degradation of insulation system. The evaluation of PD activity is the basic principle of

J. Pan et al. (eds.), *Intelligent Data Analysis and Applications*, Advances in Intelligent Systems and Computing 535, DOI 10.1007/978-3-319-48499-0_33

some methods of CC fault detection. These methods can be divided: (i) methods evaluating PD activity as a low-energy current signal [5,7,19]; (ii) methods evaluating PD activity as a voltage signal measured in CC vicinity [12].

In the case of the second approach, there is an electric field evaluated in vicinity of CC measured as a voltage signal. The high-frequency component (impulse component) within the limits (1–10 MHz) MHz of the voltage signal is obtained from this voltage signal as a so called PD-Pattern. For no-fault state of CC, the impulse component approximates to zero value and PD-Pattern was created by high-frequency disturbances of external radio transmitters. For a fault state, impulse component is generated by PD activity and shape of PD-pattern corresponds to this impulse component change.

The main goal of this paper is to design and evaluate the machine learning based detection of CCs' faults from the voltage signal using the features derived from the complex networks (CN). According to our knowledge of the state-of-the art, the analysis of PD-pulses of the PD-Pattern based on complex network features is applied for the first time. This paper covers the entire work flow of the experiment, testing and comparison to our previous models based on denoising and feature extraction.

The application of CN based features is reasonable in more ways. The natural environment where all of the signals are obtained contains high presence of noise disturbance, which creates a lot of false-hit pulses (pulses similar to PD activity but created by noise interference) inside of the signals. Such pulses are not implying any kind of damage on CC, but they could be sometimes very similar to the PD-pulses (according to application of our measuring device- Sect. 2.2). The differences are only in small number of features and in distribution of the pulses inside of the signal. These assumption leads us to apply the modeling of similarity relations between the pulses as a complex network.

2   Experiment Design

The experiment comes through several stages as it is drawn in Fig. 1. The very first stage, after all signals are stored and labeled by expert (signal obtains the class number according to appearance of PD-behavior), is the suppression of the noise from the signal (Sect. 2.1). When the most significant pulses are kept, there is acquired a matrix of pulses' feature vectors (Sect. 2.2) which will serve as dataset for complex network for the following step (Sect. 2.3). Such a complex networks is examined for feature extraction again which results into a feature vector with added class variable of the given signal (Sect. 2.4). This enables us to study the signal as a unit and not sequentially pulse by pulse. Such a dataset serves as input into the last stage of classification by Random Forest algorithm (Sect. 2.5).

The dataset of signals applied in this experiment contains only signals with higher presence of some pulses (PD or false-hits), because such signals are the hardest to recognize for previously applied models [15]. The minimal amount of pulses was considered as 150 and the total number of signals was 241.

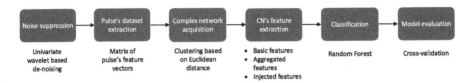

Fig. 1. Work-flow of the entire experiment.

2.1 Univariate Wavelet Based Denoising

The raw signal has to be denoised in order to suppress the major part of the noise. We use the basic univariate wavelet denoising for this purpose which is inspired by many previous studies [16]. The univariate wavelet denoising comprises of four steps:

1. Perform discrete wavelet transform (DWT) to obtain wavelet coefficients $w_{jk}(x)$ on level j from signal x.
2. Estimate the threshold T_d at decomposition level j via:

$$\sigma = 1/0.6745\,\mathrm{MAD}(|c_d|)\,,\tag{1}$$

$$T_d = \sigma\sqrt{2\log(n)}\,.\tag{2}$$

3. Perform hard threshold on both approximation and detailed coefficients.
4. Reconstruct a denoised version of the original signal from the thresholded coefficients.

The applied mother wavelet and the number of the decomposition levels was adjusted according to our previous study [15]. By application of hard thresholding, the major amount of irrelevant small pulses is suppressed and the most significant pulses remain for further analysis.

2.2 Dataset Acquisition from Pulses' Features

The pulse is the most important part of the signal and therefore it is an objective of this experiment. It is a rapid, transient change of the signal's amplitude to the lower or higher value followed by quick return to the baseline value. Each observed signal examined by this experiment contains more than 150 pulses. Some of the pulses indicate only noise behavior as corona pulses or radio waves pulses, but some of them represents pulses of PD. PD appear with well defined PD-patterns, however the application of single layered inductor (SLI) as measuring device with sampling rate 20 MS/s (expenses are not allowed to exceed $6000 according to request of operator of MV overhead lines with CC) makes impossible to recognize PD by its PD-pattern.

We considered only features extracted from each pulse which brings some relevancy according to the previous study [12] and our statistical experiments [15]. Those features are the width, height, position, closeness of symmetric pulse, and ratio between pulse's height and height of its symmetric pulse. Those feature vectors compound the dataset of pulses' features which is turned into the complex network in the next step of the experiment.

2.3   Methodology of a Complex Network Application

The feature matrix obtained from the previous step contains columns which represent pulses' features and rows representing the pulses itself. The number of pulses varied between analyzed signals and it was mostly higher than 500.

The pulses' features are normalized into range of $\langle 0,1 \rangle$ and its matrix is transformed into CN where each vector (row of the matrix) represents a node of the network and edges are created according to the given similarity function and adjusted level of threshold T.

We use the Euclidean distance as a similarity function in this experiment.

When two compared pulses are very similar, they appear very close in the given n-dimensional feature space. If the distance between two points is below the threshold value, those points are connected by edge. We are able to control the number of edges by this adjustable threshold value. The edges are undirected because euclidean distance is commutative and weighted according to the normalized value of the distance computed as follows

$$w_{i,j} = 1 - \widehat{d}_{i,j} \, . \tag{3}$$

The threshold value T depends on distances d of the given dataset and it is defined as percentage p of its normalized values.

$$T = \min(d) + ((\max(d) - \min(d))\,p) \tag{4}$$

Examples of obtained CNs are depicted in Figs. 2 and 3.

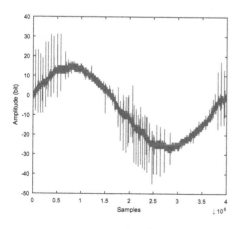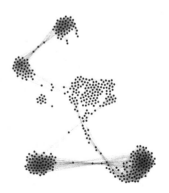

Fig. 2. Failure-free signal (left) and complex network (right) obtained from its pulses after denoising procedure

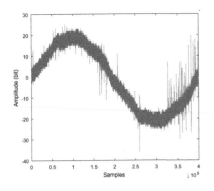
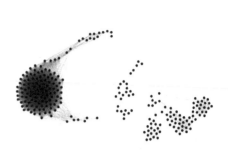

Fig. 3. Failure signal (left) (contains pulses of partial discharges) and complex network (right) obtained from its pulses after denoising procedure

2.4 Feature Extraction from Complex Networks

There are several studies of CN analysis [9,11,14], which deals with CN's features as an indicators of the system's inner behavior. In this experiment we apply several features and three different approaches of their extraction. All of the following mentioned features are contained in CN's feature vector for classification.

Basic CN's Features. In case of this experiment, a few basic features were extracted from the CNs.

The density of the network is taken as a number of edges divided by the number of all potential connections. The other feature is global clustering coefficient which was introduced and described in study of Newman [13]. It can be calculated as a number of triangles divided by number of connected triples multiplied by three.

On the other hand the local clustering coefficient was introduced much sooner [18] and it determines whether a graph is a small-world network. This coefficient is computed for each node and the averaged value of all nodes is taken as another feature for classification in this experiment.

Aggregated CN's Features. As it was concluded and compared in study of Aliakbary et al. [1] the degree distribution can be evaluated by several methods. The most widely applied techniques for comparison are the Kolmogorov-Smirnov (KS) test [3], comparison based on distribution percentiles [8], and comparison based on fitted power-law exponent [4]. The mentioned techniques are very sensitive to presence of outliers and comparisons of different sized networks, which is actually the nature of this experiment.

The feature extraction method proposed by Aliakbary [1] process the degree distribution into quantification feature vector (Eq. (12)). It contains eight Interval Degree Probability values (IDP) (Eq. (11)), which computes the probability that the selected interval I (Eq. (10)) contains the degree of randomly chosen node.

$$P_G(d) = P(D(v) = d); v \in V(G) \tag{5}$$

$$\mu_G = \sum_{d=\min_G(D(v))}^{\max_G(D(v))} d \times P_G(d) \tag{6}$$

$$\sigma_G = \sqrt{\sum_{d=\min_G(D(v))}^{\max_G(D(v))} P_G(d) \times (d - \mu_G)^2} \tag{7}$$

$$R_G(r) = \begin{cases} [\min_G(D(v)), \mu_G - \sigma_G]; & r = 1 \\ [\mu_G - \sigma_G, \mu_G]; & r = 2 \\ [\mu_G, \mu_G + \sigma_G]; & r = 3 \\ [\mu_G + \sigma_G, \max_G(D(v))]; & r = 4 \end{cases} \tag{8}$$

$$|R_G(r)| = \max(right(R_G(r)) - left(R_G(r)), 0) \tag{9}$$

$$I_G(i) = \begin{cases} [left(R_G(\lceil \frac{i}{2} \rceil)), left(R_G(\lceil \frac{i}{2} \rceil)) + \frac{|R_G(\lceil \frac{i}{2} \rceil)|}{2}]; & i \text{ is odd} \\ [left(R_G(\lceil \frac{i}{2} \rceil)) + \frac{|R_G(\lceil \frac{i}{2} \rceil)|}{2}, right(R_G(\lceil \frac{i}{2} \rceil))]; & i \text{ is even} \end{cases} \tag{10}$$

$$IDP_G(I) = P(left(I) \geq D(v) < right(I)); v \in V(G) \tag{11}$$

$$Q(G) = \langle IDP_G(I_G(i)) \rangle_{i=1..8} \tag{12}$$

Such quantification feature vectors were calculated on distribution of degree, node betweenness, and edge betweenness. This brings 24 new feature values into the final classification.

Injected CN's Features. The third approach of data-mining from the complex network is based on fundamental knowledge about the PD-pulses. Based on most relevant features of the pulse (Sect. 2.2) and supervised annotated dataset of the signals, we chose 10 most typical (visual selection of signals with lowest amount of noise and highest amount of PD interference) signals containing PD behavior.

Pulses from these signals were clustered by k-NN clustering into two clusters and the centroid points of each cluster were taken as a typical PD-pulses. Such typical pulses were injected into each of the evaluated networks with assumption, that those injected pulses (new nodes of the network) will obtain different features in fault signals and different features in failure-free signals. The features considered in this group were the degree and betweenness of the node.

2.5 Random Forest Classification

Random Decision Forest is a general title for ensemble based machine learning model which was proposed by Breiman in 2001 [2]. This model was applied with success in many machine learning studies. The core idea of the algorithm is focused on combined application of an ensemble of CART-like tree classifiers (boosting) and their learning performed on the boosted-aggregated observations (bagging).

Learning of the ensemble of trees means to train the set of trees where each of them obtain different random subset of observation and different random subset of variables. This process minimizes the correlation between the trees, which increase the robustness of the model and decrease the possible amount of over-fitting. The final classification is derived from voting mechanism where votes from all of the trees are taken into account and final class is assigned to the observation by votes of the majority of the ensemble.

The bootstrapping mechanism comes from statistic and it is also know as random sampling with replacement. This mechanism in context of RF algorithm produces balanced subset of observations for each of the tree. They are trained on resampled observations, which can handle the imbalanced problem or the problem of inability to learn some specific observations.

3   Settings and Results

The following section summarizes the setting and results of the entire experiment. The settings of the RF classifier were adjusted experimentally with respect to the previous research and those settings were kept during entire experiment (100 trees in ensemble, random subset of variables and random subset of observations for each tree to prevent over-fitting) - for all levels of threshold value.

The value of the threshold T varied. We had to examine different levels of this value to compare if the obtained CN contains enough information for best classification performance. On the other hand, in case when the value T was too high, new edges of the CN brought only noise or uncertainty.

The cross-validation method [10] was applied for testing of classification performance and the results covers calculated values of accuracy, precision, recall,

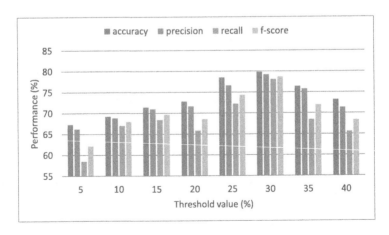

Fig. 4. Graphical comparison of classification performance based on application of different threshold levels.

Table 1. Comparison of classification performance to our previous models based on different types of de-nosing, basic feature extraction and machine learning [15].

Model	Accuracy (%)	Precision (%)	Recall (%)	F-score (%)
DWT + ANN	55.65	61.36	76.47	63.58
WPD + ANN	54.36	61.36	74.14	62.59
PSOSVD + ANN	61.85	67.73	61.21	61.85
CN's features + Random Forest	**79.82**	**79.24**	**78.13**	**78.61**

specificity, and basic F-score. The best results in our experiment were obtained on 30 % value of the threshold (see Fig. 4 and Table 1).

The performance has increased for all measured metrics compared to the previous experiment [15] (see in Table 1), which confirms the relevancy of application of complex network based features.

4 Conclusions

This paper covers the very first application of complex network based features for detection of fault behavior on covered conductors. The complex network created from extracted signal's pulses reflects valuable features for PD-pulses classification. This hypothesis was confirmed by increase of performance of the classification model compared to our previous research [15].

Model based on CN's features can be used as a supportive tool classification for signals with higher noise interference, because signals without any pulses (PD or false-hits) are automatically considered as failure-free signals and they were not studied in this paper.

The future work will cover the increased aim of data-mining on the CN's features and application of different approaches of constructing the CN. On the other hand there is still an option to study the parts of the signal separately and examine the network's dynamic for another set of relevant features for classification.

Acknowledgment. This research was conducted within the framework of the project TUCENET Sustainable Development of Centre ENET LO1404 and Students Grant Competition project reg. no. SP2016/175, SP2016/177, SP2016/128.

References

1. Aliakbary, S., Habibi, J., Movaghar, A.: Feature extraction from degree distribution for comparison and analysis of complex networks. Comput. J., bxv007 (2015)
2. Breiman, L.: Random forests. Mach. Learn. **45**(1), 5–32 (2001)
3. Clauset, A., Shalizi, C.R., Newman, M.E.: Power-law distributions in empirical data. SIAM Rev. **51**(4), 661–703 (2009)

4. Faloutsos, M., Faloutsos, P., Faloutsos, C.: On power-law relationships of the internet topology. ACM SIGCOMM Comput. Commun. Rev. **29**, 251–262 (1999)

5. Hashmi, G.M., Lehtonen, M.: Effects of rogowski coil and covered-conductor parameters on the performance of pd measurements in overhead distribution networks. Int. J. Innov. Energy Syst. Power **4**, October 2009

6. Hashmi, G., Lehtonen, M., Nordman, M.: Modeling and experimental verification of on-line pd detection in mv covered-conductor overhead networks. IEEE Trans. Dielectr. Electr. Insul. **17**(1), 167–180 (2010)

7. Hashmi, G., Lehtonen, M., Nordman, M.: Calibration of on-line partial discharge measuring system using rogowski coil in covered-conductor overhead distribution networks. IET Sci. Measur. Technol. **5**(1), 5–13 (2011)

8. Janssen, J., Hurshman, M., Kalyaniwalla, N.: Model selection for social networks using graphlets. Internet Math. **8**(4), 338–363 (2012)

9. Kaluza, P., Kölzsch, A., Gastner, M.T., Blasius, B.: The complex network of global cargo ship movements. J. Roy. Soc. Interface **7**(48), 1093–1103 (2010)

10. Kohavi, R., et al.: A study of cross-validation and bootstrap for accuracy estimation and model selection. In: IJCAI, vol. 14, pp. 1137–1145 (1995)

11. Lin, L., Wang, Q., Sadek, A.: Data mining and complex network algorithms for traffic accident analysis. Transp. Res. Record: J. Transp. Res. Board (2460), 128–136 (2014)

12. Misak, S., Pokorny, V.: Testing of a covered conductor's fault detectors. IEEE Trans. Power Delivery **PP**(99), 1 (2014)

13. Newman, M.E.: The structure and function of complex networks. SIAM Rev. **45**(2), 167–256 (2003)

14. Rubinov, M., Sporns, O.: Complex network measures of brain connectivity: uses and interpretations. Neuroimage **52**(3), 1059–1069 (2010)

15. Vantuch, T., Misak, S., Burianek, T., Jezowicz, T., Fulnecek, J.: Swarm intelligence based denoising for detection of partial discharges on covered conductors in natural environment. Adv. Electr. Electron. Eng. (in press, 2016)

16. Vidakovic, B.: Statistical Modeling by Wavelets, vol. 503. John Wiley & Sons, New York (2009)

17. Wareing, J.B.: Covered conductor systems for distribution. Technical report. 70580, EA Technology Ltd., Capenhurst Technology Park, Capenhurst, Chester, CH1 6ES, December 2005

18. Watts, D.J., Strogatz, S.H.: Collective dynamics of small-world networks. Nature **393**(6684), 440–442 (1998)

19. Zhang, W., Hou, Z., Li, H.J., Liu, C., Ma, N.: An improved technique for online pd detection on covered conductor lines. IEEE Trans. Power Delivery **29**(2), 972–973 (2014)

Author Index

© Springer International Publishing AG 2017
J. Pan et al. (eds.), *Intelligent Data Analysis and Applications*, Advances in Intelligent Systems and Computing 535, DOI 10.1007/978-3-319-48499-0

287

Printed in the United States
By Bookmasters